The
Quotable
Feynman

The Quotable Feynman

edited by
MICHELLE FEYNMAN

PRINCETON UNIVERSITY PRESS

PRINCETON AND OXFORD

Published by Princeton University Press,
41 William Street, Princeton, New Jersey 08540
99 Banbury Road, Oxford OX2 6JX

press.princeton.edu

Cover image: Bo Arrhed / Alamy Stock Photo

First paperback printing, 2025
Paperback ISBN 978-0-691-27044-9
ISBN (e-book) 978-1-400-87423-1

The Library of Congress has cataloged the cloth edition of this book as follows:

Feynman, Richard P. (Richard Phillips), 1918-1988, author.
[Quotations. Selections]
The quotable Feynman / edited by Michelle Feynman.
pages   cm
Includes index.
ISBN 978-0-691-15303-2 (hardcover : alk. paper)   1. Feynman,
Richard P. (Richard Phillips), 1918-1988—Quotations.
I. Feynman, Michelle, editor.   II. Title.
QC16.F49A25 2015
081—dc23   2015022699

British Library Cataloging-in-Publication Data is available

This book has been composed in BodoniStd-Book

Printed in the United States of America

To Ava and Marco

Contents

A Brief Note on Sources

Richard Feynman has been called the great explainer. This book of quotes characterizes my father's approach to scientific problem solving, his philosophy, and his communication style. Taken by topic, these quotes provide a richer, deeper understanding of how he thought, emphasize what he thought was important, and give shining examples of how he expressed himself.

Sources were his many published works, his personal papers that occupy 14 file drawers, and dozens of hours of recorded lectures. A number of important quotes also came from interviews that he did with Charles Weiner for an oral history project for the American Institute of Physics from 1966–1973. I, along with research assistants Anisha Cook and Janna Wennberg, who were instrumental in the book's shape over the final summer, gathered thousands of quotes and then sorted them into the 26 topics that make up this book.

While no collection of quotations drawn from written works, notes, correspondence, and lectures can fully capture my father's wide-ranging thoughts on various topics, it is my hope this compilation will provide the reader with a sense of his clarity, his humor, his unique way of looking at the world.

Michelle Feynman

Foreword

If you go into any physics department at any university in the world and ask the undergraduates which scientist they most aspire to be like, I think a majority would say "Richard Feynman." Einstein might come a close second. I would say Feynman.

Richard Feynman was one of the greatest physicists of the 20th century. His 1965 Nobel Prize, shared with Julian Schwinger and Sin-Itiro Tomonaga, was awarded for their work in developing the theory of quantum electrodynamics, which to this day stands as our most precise description of the interaction between light and matter. We wouldn't understand atoms without it. Feynman's name is most closely associated with the theory because of his introduction of Feynman diagrams. Every physicist working at CERN, or any modern particle physics laboratory, has been taught how to use Feynman diagrams. They are the foundation of our understanding of the subatomic world, allowing us to calculate what will happen when particles collide and even to predict the existence of new ones, like the Higgs Boson. I cannot imagine particle physics without them, and they probably wouldn't exist today without Feynman; I don't think anyone else would have come up with them. They are intuitively obvious after they have been explained to you, and yet you get the feeling that you'd never have invented them yourself. This was Richard Feynman's particular genius; exploring nature using a sort of internalized intuitive magic. His approach was summarized famously by his friend and colleague Hans Bethe: "There are two types of genius.

Ordinary geniuses do great things, but they leave you room to believe that you could do the same if only you worked hard enough. Then there are magicians, and you can have no idea how they do it. Feynman was a magician."

Quantum electrodynamics alone is enough to place Feynman firmly among the greats, but there are plenty of Nobel Prize–winning physicists whose names few undergraduates would even recall, let alone hero worship. The reason for the unique esteem in which Richard Feynman is held, I think, can be found in this book. It is his incisive, humble logic; razor-sharp precision deployed with humanity; wonder catalyzing discovery; a deep love of nature and a powerful desire to understand how it works. When I read his words, which should be read in a New York working-man's accent, I hear the reassuring clarity of an old engineer who's popped 'round to fix your plumbing. There is no artifice, no obfuscation, no attempt to bamboozle or self-aggrandize; just a feeling that this person will get the job done with a minimum of fuss. Richard Feynman thought about physics like that. I often quote him in my own books because I can never put my feelings about physics quite as clearly as he could. There is a beautiful interview broadcast by the BBC in 1981 called "The Pleasure of Finding Things Out," in which Feynman is asked about the possibility of discovering a "Theory of Everything" — a complete mathematical framework which describes all of Nature at the most fundamental level.

"People say to me, 'Are you looking for the ultimate laws of physics?' No, I'm not. I'm just looking to find out more about the world, and if it turns out there is a simple ultimate law which explains everything, so be it; that would be very nice to discover. If it turns out it's like an onion with millions of layers and we're just sick and tired of looking at the layers, then that's the way it is! And therefore when we go to investigate we shouldn't pre-decide what it is we are trying to do except to find out more about it. . . . My interest in science is to simply find out more about the world."

That is, for me, a perfect description of science. Take delight in the small, rather than adopting futile intellectual postures in the face of the intricate infinite delicacy of nature and you might make a little progress. Reading his words, you will hear that message over and over again. I am a simple man, and I like to think carefully about simple things. That's a real physicist talking.

This direct simplicity certainly didn't preclude lyricism when it came to describing the process of doing science. Some of my favorite quotes are those that reveal Feynman's inner picture of the scientific endeavor: "We live in a heroic, a unique and wonderful age of excitement. It's going to be looked at with great jealousy in the ages to come. How would it have been to live in the time when they were discovering the fundamental laws?" This is suffused with, to use a cliché often hurled at scientists, child-like wonder, and Feynman was absolutely comfortable with such a double-edged compliment. "I hate adults."

Feynman was also a polemicist, deploying his deliberate linguistic clarity to powerful effect whenever he got the chance. I can think of few physicists who would write, in an introductory lecture on electromagnetism: "From a long view of the history of mankind, seen from, say, ten thousand years from now, there can be little doubt that the most significant event of the 19th century will be judged as Maxwell's discovery of the laws of electrodynamics. The American Civil War will pale into provincial insignificance in comparison with this important scientific event of the same decade." I love this. I read it as an entirely appropriate swipe at the futility of provincial wars, prefiguring Carl Sagan's *Pale Blue Dot*, the great lament for parochial human stupidity written three decades later: "The Earth is a very small stage in a vast cosmic arena. Think of the rivers of blood spilled by all those generals and emperors so that in glory and triumph they could become the momentary masters of a fraction of a dot."

In these pages, then, you will catch a glimpse of genius, read the thoughts of one of history's great polymaths and one of the greatest scientists who ever lived. I hope you will learn a lot, as I have, and perhaps emerge with a little additional curiosity, wonder, humility, and respect for the natural world. Having said that, "I don't wanna take this stuff seriously. I think we should just have fun imagining it and not worry about it."

Brian Cox
Professor of Particle Physics
The University of Manchester

Reflections on
Richard Feynman

There seems to be an insatiable public appetite to feel a closeness with the giant mind and personality of Richard Feynman. This curiosity transcends generations, crosses disciplines and cultures. Over a quarter of a century after his passing, he remains alive in the public consciousness, his books are still in print, his legendary lectures are online, scientists are continuing to grapple and tangle with many of the theories he put forth decades ago.

So where does the longevity of his aura come from? I can only offer a sliver of a snapshot.

Over three decades ago, I used to see Richard backstage at concerts. He came not because he liked cello playing particularly, but because his beloved young daughter, Michelle, played, and of course, what doting father wouldn't want to please his daughter? Sometimes we would pass time bantering about what truth is in science and art, and he would always say, "In science you have to prove it." And then he would regale us with stories of his adventures playing the bongo drums. Once we went to the house and he showed us his beautiful drawings of the human figure. He talked about how his desire to go to Tuva came about from playing a geography game. He was always energetic, attentive, and present.

One of my heroes, as I was growing up, was the great cellist Pablo Casals. I was particularly impressed when he remarked that he was a human being first, a musician second, and a cellist third. I was similarly taken when I read one of Richard's quotes: "You cannot develop a personality with physics alone; the rest of your life must be worked in."

Therein lies a clue to the longevity of Richard Feynman. Yes, he was one of the all-time great physicists, but he also paid attention to life and love, to his children, to his family, to the sensuality of the human figure, to the primal complexities of drumming, to his entire environment. While he paid close attention to problems we face and generate, he also knew that humans are a subset of nature, and nature held for him the greatest fascination — for the imagination of nature is far, far greater than the imagination of man, and nature guards her secrets jealously.

Thus, for him, it was worth the years of work to extract some of those secrets in order to pass them on in the most direct and understandable form to the rest of us. Because he worked his whole life into his personality, we could identify with his humanness and therefore stay with him as he took us on the most spectacular journey of all, the never-ending quest to understand everything.

Surely you must still be joking, Mr. Feynman!

Yo-Yo Ma
Cellist

Preface

My Quotable Father

I recently came across a few of my notebooks from high school and college and found scribbles from my time in the classroom. By turns funny, uplifting, heartbreaking, and occasionally annoying, those excerpts from my life of years ago remind me of a particular moment or a class much more quickly than do my actual class notes — there is something very powerful about a concise statement. And so I have always loved quotes.

One of my favorite quotes from my father came in a birthday card he gave me when I turned 18: "Go forth!" he wrote. When I read those words, I remember my reaction — pride and excitement, mixed with some apprehension. I also recall that he usually didn't bother himself with the trivialities of birthday cards; he left that to my mother, which made this particular quote that much more poignant.

My father left me his words. From them I remember both his outlook on life and his voice, positive and clear. He was someone who didn't worry about small problems. His advice here — leave it; let it go — is inspiring:

> And we all do stupid things, and we know some people do more than others, but there is no use in trying to check who does the most.

I'm often asked what kind of father he was, and although I occasionally teased him for not remembering what I thought were pertinent details of my life (age, school, etc.), he was always happy and ready to spend time with me. He might have had a reputation for not suffering fools gladly, but I remember him as a funny, energetic, kind, playful, patient man. I recall his wise advice about finding a career, very similar to a letter he wrote to a high school student in 1984:

> It is wonderful if you can find something you love to do in your youth which is big enough to sustain your interest through all your adult life. Because, whatever it is, if you do it well enough (and you will, if you truly love it), people will pay you to do what you want to do anyway.

Whenever I read his words, I hear his voice. It still makes me laugh to think that he used the phrase "cooked up" in his Nobel Prize lecture:

> I never used all that machinery which I had cooked up to solve a single relativistic problem.

I am charmed by my father's use of expressions that are no longer in rotation, and I am mesmerized by his distinct rhythm of speaking. I remember a few archaic expressions from when I was growing up — the refrigerator was the "icebox," for goodness' sake! Or "They go all around Robin Hood's barn to come around here" to describe a round-about path.

I know he was fascinated by Las Vegas. He incorporated stories from visits there in his lectures. I really love the following excerpt because he was such an expert at breaking grammatical rules. The rhythm really sets the scene:

Example. I'm in Las Vegas, suppose. And I meet a mindreader, or, let's say, a man who claims not to be a mindreader, but more technically speaking to have the ability of telekinesis, which means that he can influence the way things behave by pure thought. This fellow comes to me, and he says, "I will demonstrate this to you. We will stand at the roulette wheel and I will tell you ahead of time whether it is going to be black or red on every shot."

I believe, say, before I begin, it doesn't make any difference what number you choose for this. I happen to be prejudiced against mindreaders from experience in nature, in physics. I don't see, if I believe that man is made out of atoms and if I know all of the — most of the — ways atoms interact with each other, any direct way in which the machinations in the mind can affect the ball. So from other experience and general knowledge, I have a strong prejudice against mind readers. Million to one.

Now we begin. The mind reader says it's going to be black. It's black. The mind reader says it's going to be red. It's red. Do I believe in mindreaders? No. It could happen. The mindreader says it's going to be red. It's red. Sweat. I'm about to learn something . . .

It was also revealing to see how he spent his leisure time. Was I surprised to see quotes that revealed his penchant for always working on physics? Not really. It was a switch he couldn't turn off. I remember he always seemed to be thinking about physics. When he didn't have a pad of paper, it was common to find equations on scraps of paper — even margins of the newspaper. Even when he was very young, he recalled, he was annoyed at being pulled away from working.

I didn't get to do as much as I wanted to, because my mother kept putting me out all the time, to play.

That one really made me laugh. These next two exemplify his open, honest attitude and how he loved nothing better than thinking about physics:

> It is a nice problem, however, and I often spend time on airplanes trying to figure it out. I haven't cracked it yet.
>
> It must have been a day or so later when I was lying in bed thinking about these things that I imagined what would happen if I wanted to calculate the wave function at a finite interval later.

I've always been impressed by his humility:

> I suffer from the disease that all professors suffer from — that is, there never seems to be enough time, and I invented more problems than undoubtedly we'll be able to do, and therefore I've tried to speed things up by writing some things on the board beforehand, with the illusion that every professor has: that if he talks about more things, he'll teach more things. Of course, there's only a finite rate at which material can be absorbed by the human mind, yet we disregard that phenomenon, and in spite of it we go too fast.

I remember so much advice from my father: Think about a math problem to see whether your answer makes sense. Work to be direct and honest in communication. Aim to be friendly and kind. Realize that life is an exciting adventure. Find something to do that you love. Work hard. And always, *always* remember your sense of humor! I have not always been able to laugh at myself, but it was something at which my father excelled. I remember his once trudging through an airport, luggage in tow, laboring to make a connecting flight home. He muttered to himself, "Flying is for the birds!" — and, realizing just how true that was, laughed and regained his equanimity. Of all the advice he dispensed, perhaps his exhortation to remember one's sense of humor is the most crucial. Doing so has helped me cope with difficult moments in my own life.

Strangely enough, I came across one of my father's quotations about *his* father that sums up how I feel about *my* father:

> Not having experience with many fathers, I didn't realize how remarkable he was.

I will be forever grateful to my father for sharing his life with me — from the charming, silly, playful moments to the serious.

Michelle Feynman, 2014

Chronology

The Quotable Feynman

Youth

I didn't get to do as much as I wanted to, because my mother kept putting me out all the time, to play.

– *Surely You're Joking, Mr. Feynman!*, p. 17

When I was a kid, I had this notion that you could take the importance of the problem and multiply it by your chance of solving it. You know how a technically minded kid is, he likes the idea of optimizing everything anyway, if you can get the right combination of those factors, you don't spend your life getting nowhere with a profound problem, or solving lots of small problems that others could do just as well.

– *Omni* interview, February 1979

Don't despair of standard dull textbooks. Just close the book once in a while and think what they just said in your own terms as a revelation of the spirit and wonder of nature. The books give you facts but your imagination can supply life. My father taught me how to do that when I was a little boy on his knee, and he read the *Encyclopaedia Britannica* to me!

– Letter to Rodney C. Lewis, August 1981 (*Perfectly Reasonable Deviations from the Beaten Track*, pp. 332–333)

I went to take the calculus book out, and the teacher — sorry, the librarian — said, "Child, you can't take this book out. Why are you taking this book out?" I said, "It's for my father." And so I took it home, and I tried to learn a little bit. My father looked at the first few paragraphs and couldn't understand it, and this was rather a shock to me — a little bit of a shock, I remember. It was the first time I realized that I could understand what he couldn't understand.

– Interview with Charles Weiner, March 4, 1966 (Niels Bohr Library and Archives with the Center for the History of Physics)

I learned very early the difference between knowing the name of something and knowing something.

– *What Do You Care What Other People Think?*, p. 14

When I was a child and found out Santa Claus wasn't real, I wasn't upset. Rather, I was relieved that there was a much simpler phenomenon to explain how so many children all over the world got presents on the same night.

– *Los Angeles Times*, November 27, 1994

When I was young, what I call the laboratory was just a place to fiddle around, make radios and gadgets and photocells and whatnot. I was very shocked when I discovered what they call a laboratory in a university. That's a place where you are supposed to measure something very seriously. I never measured a damn thing in my laboratory.

– Future for Science interview

[On his first talk:] I remember getting up to talk, and there were these great men in the audience and it was frightening. And I can still see my own hands as I pulled out the papers from the envelope that I had them in. They were shaking. As soon as I got the paper out and started to talk, something happened to me which has always happened since and which is a wonderful thing. If I'm talking physics, I love the thing. I think only about physics, I don't worry where I am; I don't worry about anything, and everything went very easily.

– Future for Science interview

The moment I realized that I was now working on something new was when I read something about quantum electrodynamics at the time, and I read a book, and I learned about it. For example, I read Dirac's book, and they had these problems that nobody knew how to solve. I couldn't understand the book very well because I wasn't up to it, but at the last paragraph at the very end of the book, it said, "Some new ideas are here needed!" And so there I was! Some new ideas were there needed, so I started to think of new ideas.

– Interview with Yorkshire Television program, "Take the World from Another Point of View," 1972

[To one of his former high school teachers:] Another thing that I remember as being very important to me was the time when you called me down after class and said, "You make too much noise in class." Then you went on to say that you understood the reason, that it was that the class was entirely too boring. Then you pulled out a book from behind you and said, "Here, you read this, take it up to the back of the room, sit all alone and study this; when you know everything that is in it, you can talk again." And so, in my physics class I paid no attention to what was going on but only studied Woods' *Advanced Calculus* up in the back of the room. It was there that I learned about gamma functions, elliptic functions, and differentiating under an integral sign. A trick at which I became an expert.

– Letter to Abram Bader, November 1965 (*Perfectly Reasonable Deviations from the Beaten Track*, pp. 176–177)

[CBS] asked me what I thought of the New York School System, and I said that I am only good in physics and I do not know the

New York School System except for the particular school that I went to thirty years ago. I thought that my high school was very good. There was a great variety of science courses offered for those times — advanced math, physics, chemistry, and biology. Several teachers gave me direct encouragement, good advice, and taught me special things outside the regular courses. I had a good time in high school.

– Letter to Miriam Cohen, November 1965

[To his aunt:] It is good to hear from someone who has known me for so long. You have gone through all the stages with mother, from ruined linen towels to mom's worrying about whether I would blow up the house with my laboratory.

– Letter to Jesse M. Davidson, December 1965 (*Perfectly Reasonable Deviations from the Beaten Track*, p. 181)

[On his father:] He was rational; he liked the rational mind and things that could be understood by thinking.

– Interview with Charles Weiner, March 4, 1966 (Niels Bohr Library and Archives with the Center for the History of Physics)

When I got to kindergarten, which was much later — I was six years old — they had a thing in those days which was "weaving." They had a kind of colored paper — square paper with quarter-inch slots made parallel. And you have quarter-inch strips of paper. One was the weft and the other was the warp. You're supposed to weave it and make designs that were regular and interesting. And apparently that's extremely difficult for a child. I was especially commented on; the teacher was very excited and surprised. I made elaborate patterns — correctly, without any difficulty, whereas it was so difficult for most of the children that they don't do that in kindergarten anymore.

– Interview with Charles Weiner, March 4, 1966 (Niels Bohr Library and Archives with the Center for the History of Physics)

My father would often take me to the Museum of Natural History — that was a great place. We would look at the dinosaur bones and stuff like that — it was great!

– Interview with Charles Weiner, March 4, 1966 (Niels Bohr Library and Archives with the Center for the History of Physics)

[On his father describing glaciers:] He understood! The thing that was very important about my father is not the facts but the process — the meaning of everything. How we find out; what is the consequence of finding such a rock? With a vivid description of the ice, which is probably not exactly right! Perhaps the speed was not ten inches a year but ten feet a year — I never knew; he never knew. But he would describe anyway, in a vivid way, and always with some kind of lesson about it. Like, "How do you think we find these things out?"

– Interview with Charles Weiner, March 4, 1966 (Niels Bohr Library and Archives with the Center for the History of Physics)

[On his sister, also a physicist:] She would hear us talking, and she would ask me, and I would explain it to her. It wasn't so direct in her case.

> – Interview with Charles Weiner, March 4, 1966 (Niels Bohr Library and Archives with the Center for the History of Physics)

I was always very upset if something went bad or if I was bad — I always tried to be a good boy.

> – Interview with Charles Weiner, March 4, 1966 (Niels Bohr Library and Archives with the Center for the History of Physics)

Arithmetic was very easy; it was too easy. For instance, when I was ten or eleven, one day I was called from a class to a previous class that I had been in by a previous teacher to explain to the class how to do subtraction. I had "invented," (they claim) a better way of doing subtraction than they were using that she liked. She had forgotten it, in the meantime, so I was called from class to explain it to her.

> – Interview with Charles Weiner, March 4, 1966 (Niels Bohr Library and Archives with the Center for the History of Physics)

[On his friend Bernard Walker:] I had a friend who was as interested in science as I was, so we did much together — I was about twelve. We studied together, we'd argue together, we did chemistry experiments.

> – Interview with Charles Weiner, March 4, 1966 (Niels Bohr Library and Archives with the Center for the History of Physics)

I was not good at athletics. This always bothered me — I felt like a sissy because I couldn't play baseball. It was to me, at a childish age, a very serious business. I had trouble learning how to ride a bicycle . . . . Every once in awhile, I would get kicked out of the group. We had a hut, and each time I was kicked out of the group, I would invent something, like a periscope for the hut or a design for a second story or something.

> – Interview with Charles Weiner, March 4, 1966 (Niels Bohr
> Library and Archives with the Center for the History of Physics)

We put sodium ferrocyanide — sodium ferrocyanide? — or something, in the towels, and another substance, an iron salt, probably alum, in the soap. When they come together, they make blue ink. So we were supposed to fool my mother, you see. She would wash her hands, and then when she dried them, her hands would turn blue. But we didn't think the towel would turn blue. This was all in the Cedarhurst era. Anyway, she was horrified. The screams of "My good linen towels!" But she was always cooperative. She never was afraid of those experiments.

> – Interview with Charles Weiner, March 4, 1966 (Niels Bohr
> Library and Archives with the Center for the History of Physics)

[On boiling water:] I remember using the developing trays, which were waxed, so that they were insulated, putting water in them, and boiling it — and watching the most beautiful phenomenon at the end, when all the water boils away, and the last bit of water, it's dry, is making sparks, because it's breaking the circuit. And the sparks move around, because it breaks here, but the water flows, you see, and it flows here and connects, and then it makes another spark here, and finally, these lines of salt, and beautiful yellow and blue sparks! It's a very beautiful thing. In fact, now that you remind me, I think I'll have to set one up and see what it looks like, after all these years. I used to boil water all the time with this thing.

— Interview with Charles Weiner, March 4, 1966 (Niels Bohr Library and Archives with the Center for the History of Physics)

I had lots of trouble, because I remember, my friend and I — the man drew on the blackboard (I still remember, you know, he's going to explain how a projection system works, you know, the projector that makes pictures on the wall) — so he drew a light bulb, and he draws a lens and so on to explain. Then he draws lines coming out of the light bulb parallel, the rays of light going parallel to each other. So, I don't remember whether it was I or my friend, but one of us said, "But that can't be right. The rays come out from the filament radially, in all directions." I don't know if I used the word "radially," but anyway, we explained. He turned around and said, "I say they go parallel, so they go parallel!" Well, this didn't sit well with us, because I knew, certainly, that no matter what he said, the rays didn't go parallel.

— Interview with Charles Weiner, March 4, 1966 (Niels Bohr Library and Archives with the Center for the History of Physics)

[On the Great Depression:] There was also the attitude that you should do something, work — you know, the idea that to hang around and do nothing was somehow . . . . There was a feeling of some sort of responsibility to earn money. I can't explain it.

  – Interview with Charles Weiner, March 4, 1966 (Niels Bohr Library and Archives with the Center for the History of Physics)

I always kept up this ability to work very quickly with the mathematics, so as to get rid of the homework.

  – Interview with Charles Weiner, March 5, 1966 (Niels Bohr Library and Archives with the Center for the History of Physics)

I don't know much about the "general theory of intelligence," but I do remember when I was young I was very one-sided. It was science and math and no humanities (except for falling in love with a wonderful intelligent lover of piano, poetry writing, etc.).

  – Letter to Dr. William L. McConnell, March 1975 (*Perfectly Reasonable Deviations from the Beaten Track*, p. 281)

I was inspired by the remarks in these books [Heitler and Dirac]; not by the parts in which everything was proved and demonstrated carefully and calculated, because I couldn't understand those very well. At the young age what I could understand were the remarks about the fact that this doesn't make any sense, and the last sentence of the book of Dirac I can still remember, "It seems that some essentially new physical ideas are here needed." So, I had this as a challenge and an inspiration. I also had a personal feeling, that since they didn't get a satisfactory answer to the problem I wanted to solve, I don't have to pay a lot of attention to what they did do.

– From *Nobel Lectures, Physics 1963–1970*, Elsevier Publishing Company, Amsterdam, 1972

At the age of thirteen I was converted to non-Jewish religious views.

– Letter to Tina Levitan, January 1967 (*Perfectly Reasonable Deviations from the Beaten Track*, p. 234)

You see, what happened to me, what happened to the rest of us is we started for a good reason, but then we're working very hard to do something, and to accomplish it, it's a pleasure, it's excitement.

– UCSB talk, "Los Alamos from Below," February 1975

Family

I have a nice cozy house with a good family in it.

– Letter to professors Gilberto Bernadini and Luigi A. Radicati
(*Perfectly Reasonable Deviations from the Beaten Track*,
p. 209)

I was happily surprised by your present of the beautiful photograph, which is now in my office. Thank you very much. When I came home with it and showed it to my son (who is twelve years old) he was delighted with it. I asked him what it was — after a few moments he said, "probably a diffraction pattern from a laser from a regular pattern of square holes." I could have killed him! I was afraid to ask him for the focal length of the lens used!

   – Letter to Sheila Sorensen, October 1974

Another thing my father told me — and I can't quite explain it, because it was more an emotion than a telling — was that the ratio of the circumference to the diameter of all the circles was always the same, no matter what the size. That didn't seem too unobvious, but the ratio had some marvelous property. There was a mystery about this number that I didn't quite understand as a youth, but this was a great thing, and the result was that I looked for pi everywhere.

   – National Science Teachers Association Fourteenth Convention
   lecture, "What Is Science?" April 1966

[On his son:] He's a lot like me, so at least I've passed on this idea that everything is interesting to at least one other person. Of course, I don't know if that's a good thing or not, you see?

   – *Omni* interview, February 1979

[Advice to a father about a son:] The two of you — father and son — should take walks in the evening and talk (without purpose or routes) about this and that. Because his father is a wise man, and the son I think is wise too for they have the same opinions I had when I was a father and when I was a son too. These don't exactly agree, of course, but the deeper wisdom of the older man will grow out of the concentrated energetic attention of the younger. Patience.

– Letter to Mr. V. A. Van Der Hyde, July 1986 (*Perfectly Reasonable Deviations from the Beaten Track*, p. 415)

[On his son:] He talks a blue streak. He could win a Nobel Prize for talking.

– *South Shore Record*, October 28, 1965

I was surprised to read your comment about my meeting the press that you did not mention how cute and wonderful my little boy looks. Could that be modesty?

– Letter to Dr. Richard Pettit, MD, the doctor who delivered son Carl, November 1965 (*Perfectly Reasonable Deviations from the Beaten Track*, p. 186)

[Advice to a father:] Do not be too mad at Mike for his C in physics. I got a C in English Literature. Maybe I never would have received a prize in physics if I had been better in English.

– Letter to Arnold Phillips, November 1965 (*Perfectly Reasonable Deviations from the Beaten Track*, p. 185)

[Advice to a father about a son:] Let him go, let him get all distorted studying what interests him the most as much as he wants. True, our school system will grade him poorly — but he will make out. Far better than knowing only a little about a lot of things.

– Letter to Mr. V. A. Van Der Hyde, July 1986 (*Perfectly Reasonable Deviations from the Beaten Track*, p. 415)

I have worked on innumerable problems that you would call humble but which I enjoyed and felt very good about because I sometimes could partially succeed.

– Letter to Koichi Mano, February 1966 (*Perfectly Reasonable Deviations from the Beaten Track*, p. 201)

We bought a new camping van that we drive around. We like to camp out in the desert and so on and it's decorated all around with these diagrams.

– BBC interview, "Scientifically Speaking," April 1976

[On his father:] When I got older, he'd take me for walks in the woods and show me the animals and birds and so on. He'd tell me about the stars and the atoms and everything else. He'd tell me what it was about them that was so interesting. He had an attitude about the world and the way to look at it which I found was deeply scientific for a man who had no direct scientific training.

– Future for Science interview

My father had taught me to worship pi; to be awe-inspired by pi. He loved pi because there was such a strange ratio and it was such a simple thing with a circle.

– Future for Science interview

[On his father:] He had ways of looking at things. He used to say that suppose we were Martians and we came down to earth and then we would see these strange creatures doing things, what would we think? For instance, he would say, to take an example, suppose that we never went to sleep. We're Martians, but we have a consciousness that works all the time and finds these creatures who for eight hours a day stop and close their eyes and become more or less inert. We'd have an interesting question to ask them. We'd say "How does it feel to do that all the time? What happens to your ideas? You're running along very well, you're thinking clearly — and what happens? Do they suddenly stop? Or do they go more and more slowly and stop or exactly how do you turn off thoughts? Then later I thought about that a lot, and I did experiments in college to try and find out the answer to that — what happened to your thoughts when you went to sleep.

– Future for Science interview

Not having experience with many fathers, I didn't realize how remarkable he was. How did he learn the deep principles of science and the love of it, what's behind it, and why it's worth doing?

– *What Do You Care What Other People Think?*, p. 14

[On his father:] Before I was born, he said to my mother that this boy is going to be a scientist. You can't say things like that in front of women's lib these days, but that is what they said in those days. But he never told me to be a scientist. The way he did it when I was very small, he'd tell me things.

— Future for Science interview

Autobiographical

Some people say, "How can you live without knowing?" I do not know what they mean. I always live without knowing. That is easy. How you get to know is what I want to know.

– "The Uncertainty of Science," John Danz Lecture Series, 1963

I remember the work I did to get a real degree at Princeton and the guys on the same platform receiving honorary degrees without work — and felt an "honorary degree" was a debasement of the idea of a "degree which confirms certain work has been accomplished." It is like giving an "honorary electricians license." I swore then that if by chance I was ever offered one I would not accept it.

> – Letter to Dr. George W. Beadle (*Perfectly Reasonable Deviations from the Beaten Track*, p. 233)

The work that I do, none of it is secret. It is internationally known; it's transmitted by letter, back and forth, and by magazines all over the world.

> – Interview for Viewpoint

Please appreciate that it is not a matter of deep principle to me, so that if my resignation causes you any major difficulties, feel free to disregard my preferences. But your flock will contain one very peculiar, sad and reluctant bird.

> – Letter to Dr. Detlev W. Bronk and the National Academy of Sciences, August 1961 (*Perfectly Reasonable Deviations from the Beaten Track*, p. 109)

It is most interesting and revealing to pick up the morning newspaper and read about oneself, especially when it is so well written and complimentary.

> – Letter to Dr. Irving S. Bengelsdorf (*Los Angeles Times*), April 1967

I'd average about 20 minutes of doing nothing, and then I'd open it, you see; well, I opened it right away to see that everything was all right and then I'd sit there for 20 minutes to give myself a good reputation that it wasn't too easy, there was no trick to it, no trick to it. And then I'd come out, you know, sweating a bit, and say "It's open. There you are." And so forth.

– UCSB talk, "Los Alamos from Below," February 1975, regarding his hobby of safecracking

In order to work hard on something, you have to get yourself believing that the answer's over there, so you'll dig hard there, right? So you temporarily prejudice or predispose yourself, but all the time, in the back of your mind, you're laughing. Forget what you hear about science without prejudice. Here, in an interview, talking about the Big Bang, I have no prejudice, but when I'm working, I have a lot of them.

– *Omni* interview, February 1979

[After discussing calculating the position of Venus:] I don't know about the philosophy of Mayans — we have very little information due to the efficiency of the Spanish conquistadores and, well, mostly their priests, who burned all the books. They had hundreds of thousands of books and there's three left. And one of them has this Venus calculation, and that's how we know about that. Just imagine our civilization reduced to three books — the particular ones left by accident, which ones?

– "QED: Photons — Corpuscles of Light," Sir Douglas Robb Lectures, University of Auckland, 1979

I am a stick-in-the-pleasant-mud type of guy.

> – Letter to professors Gilberto Bernardini and Luigi A. Radicati, February 1966 (*Perfectly Reasonable Deviations from the Beaten Track*, p. 209)

Old friends should tell each other what they are doing.

> – Letter to Morrie Jacobs, November 1965 (*Perfectly Reasonable Deviations from the Beaten Track*, p. 179)

But whatever way I did it, I wouldn't give a damn. I know I didn't care to find out what way you had to do it, because it seemed to me, if I did it, I did it.

> – Interview with Charles Weiner, March 4, 1966 (Niels Bohr Library and Archives with the Center for the History of Physics)

It is interesting that human relationships, if there is an independent way of judging truth, can become unargumentative.

> – "The Uncertainty of Science," John Danz Lecture Series, 1963

But I never thought of theoretical physics as being a field that's split-able, in those days — just anything in theory. Because, you see, my mathematical talents had finally overcome my experimental talents. I used to play around, but I played around less and less and did more and more analysis, mathematical, like these theorems and these papers, so I was becoming theoretical instead of experimental.

> – Interview with Charles Weiner, March 5, 1966 (Niels Bohr Library and Archives with the Center for the History of Physics)

I think it is very important to resist accepting too much of unimportant details and inanities. Some wisdom and skill is required to avoid throwing the baby with the bath. But, otherwise, we have too much to think about and cannot concentrate our small minds on the problems.

– Letter to Mr. Jim Barclay, January 1967

It's hard to work in secrecy; there's a kind of schizophrenia involved in the sense that you have to remember what it is that you know that you're not allowed to say. And what happens is that you lean over backwards to avoid saying things until you become inarticulate on certain subjects, because if you start to talk about a subject which you heard something secret, you're afraid to say anything about it for fear that maybe that was one of the things that was secret. So I don't like secrets.

– Interview for Viewpoint

[On philosophers:] It isn't the philosophy that gets me, it's the pomposity. If they'd just laugh at themselves!

– *Omni* interview, February 1979 (*The Pleasure of Finding Things Out*, p. 195)

I find it psychologically very distasteful to judge people's "merit."

– Letter to Dr. Detlev W. Bronk and the National Academy of Sciences, August 1961 (*Perfectly Reasonable Deviations from the Beaten Track*, p. 108)

My lament was that a kind of intense beauty that I see given to me by science, is seen by so few others.

> – Letter to Mrs. Robert Weiner, October 1967 (*Perfectly Reasonable Deviations from the Beaten Track*, p. 248)

I suspect that there are plenty of ingenious people able to deal, as well as I, with the problems that you mention, who are at present employed as anything from manager to criminal.

> – Letter to Jarrold R. Zacharias, Laboratory for Nuclear Science and Engineering (*Perfectly Reasonable Deviations from the Beaten Track*, p. 82)

It was such a shock to me to see that a committee of men could present a whole lot of ideas, each one thinking of a new facet, while remembering what the other fellow said, so that, at the end, the decision is made as to which idea was the best — summing it all up — without having to say it three times. So that was a shock. These were very great men indeed.

> – On his experience with the Manhattan Project, "Los Alamos from Below," 1975

I enjoyed leafing through your magazine *Science and Children*, but find I must decline your invitation to write an article for it. I find that my talents are much better in the line of dancing and being an escort than in writing articles for a magazine of that type.

> – Letter to Diane Ruth (*Science and Children*), June 1966

So I made my decision, not to be a member of an honorary society, if it's an honorary society only. If this damned thing did anything, it would be all right.

– Interview with Charles Weiner, June 28, 1966 (Niels Bohr Library and Archives with the Center for the History of Physics)

[On *Surely You're Joking, Mr. Feynman!*:] I was telling all these stories to a friend. There was no idea that I was telling them to anybody else, so there was no correcting or worrying about how stupid it looked or how clever or whether I was egotistical or an idiot in the story. It didn't make any difference. I'd tell it the way it happened to me. Then Ralph got the idea to write them down, so he arranged them a little bit and kept them sounding like me.

– "The Remarkable Dr. Feynman," *Los Angeles Times Magazine*, April 20, 1986

Now that I am burned out and I'll never accomplish anything, I've got this nice position at the university teaching classes which I rather enjoy, and just like I like reading *Arabian Nights* for pleasure, I'm going to play with physics, whenever I want to, without worrying about any importance whatsoever.

– *Surely You're Joking, Mr. Feynman!*, p. 173

I'm not much good at this show business, but we'll see what happens.

– Audio recording of Feynman Lectures on Physics, Lecture 12, November 7, 1961

I knew I'd be a scientist somehow.

> – Interview with Charles Weiner, March 4, 1966 (Niels Bohr
> Library and Archives with the Center for the History of
> Physics)

Things aren't really so tough; it is just fun having crazy
adventures. I get a kick out of life and so sometimes odd things
befall me.

> – Letter to Mary Bowers, November 1960

Mulaika would rejoice in knowing that I am sitting here with
one of the most beautiful "shiners," I, at least have ever seen.
Apparently the bars in Buffalo are tougher than even the chili
and sailors near the Brooklyn Navy Yard.

> – Letter to Bert and Mulaika Corben, 1948

Probably, it was just the kind of time everyone has in California,
but it seemed to me to be especially good.

> – Letter to Professor E. O. Lawrence, July 1947 (*Perfectly
> Reasonable Deviations from the Beaten Track*, p. 76)

The Brazilians were impressed with my Portuguese, in fact the
lectures were given in the language which they referred to as
Feynman's Portuguese.

> – Letter to Professor Joe Keller, September 1949

[Regarding a request to do a talk:] Oh heck, we know each other well enough, what I mean is I don't feel like doing it.

– Letter to Dr. Victor F. Weisskopf, April 1962

The first morning I drove in was tremendously impressive; the beauty of the scenery, for a person from the east who didn't travel much, was sensational. There are the great cliffs; you've probably seen the pictures, I won't go into much detail. These things were high on the mesa and you'd come up from below and see these great cliffs and we were very surprised. The most impressive thing to me was that as I was going up, I said that maybe there were Indians even living here, and the guy who was driving the car just stopped; he stopped the car and walked around the corner and there were Indian caves that you could inspect.

– UCSB talk, "Los Alamos from Below," February 1975

I go down to the dormitory place to get rooms assigned, and they say you can pick your room now. I tried to pick one; you know what I did? I looked to see where the girl's dormitory was and I picked one that you could look out across. Later I discovered a big tree was growing right in front of it.

– UCSB talk, "Los Alamos from Below," February 1975

Well, the bachelors, and the bachelor girls, the people who lived in the dormitory, felt they had to have a faction because a new rule had been promulgated — no women in the men's dorm, for instance. Well, this is absolutely ridiculous. All grown people of course. What kind of nonsense? So we had to have political action.

– UCSB talk, "Los Alamos from Below," February 1975

The trouble with playing a trick on a highly intelligent man like Mr. Teller is the time it takes him to figure out from the moment that he sees there is something wrong to till he understands exactly what happened is too damn small to give you any pleasure!

– UCSB talk, "Los Alamos from Below," February 1975

I can't forget Hawaii. I look at my old face mask and gloves and long to be with the fish of Hanauma Bay again.

– Letter to Dr. San Fu Tuan (University of Hawaii), September 1973

I met some very great men beside the men on the evaluation committee, the men I met in Los Alamos. And there are so many of them that it's one of my great experiences in life to have met all these wonderful physicists.

– UCSB talk, "Los Alamos from Below," February 1975

I don't know anything, but I do know that everything is interesting if you go into it deeply enough.

– *Omni* interview, February 1979

There is a difference between the name of the thing and what goes on.

– National Science Teachers Association Fourteenth Convention lecture, "What Is Science?" April 1966

Perhaps it is just that I enjoy being peculiar.

– Letter to Dr. Detlev W. Bronk and the National Academy of Sciences, August 1961 (*Perfectly Reasonable Deviations from the Beaten Track*, p. 108)

It is my policy to keep up the standard of my own published work.

– Letter to Alladi Ramakrishnan, January 1962 (*Perfectly Reasonable Deviations from the Beaten Track*, p. 130)

Please be assured that I am not interested in propaganda or persuading people. We agree that we should be interested only in what is right in science.

– Letter to F. Harrison Stamper, February 1962

Words can be meaningless. If they are used in such a way that no sharp conclusions can be drawn, as in my example of "oomph," then the proposition they state is almost meaningless.

– "The Uncertainty of Science," John Danz Lecture Series, 1963

One might say that, in case you are beginning to believe that some of the things I said before are true because I am a scientist and according to the brochure that you get I won some awards and so forth, instead of looking at the ideas themselves and judging them directly. In other words, you see, you have some feeling toward authority. I will get rid of that tonight. I dedicate this lecture to showing what ridiculous conclusions and rare statements such a man as myself can make. I wish, therefore, to destroy and image of authority that has previously been generated.

> – "The Unscientific Age," John Danz Lecture Series, 1963

Among the many things I know very little about, one is what one should do to prepare oneself to be a theoretical physicist.

> – Letter to Eric W. Leuliette, September 1984 (*Perfectly Reasonable Deviations from the Beaten Track*, p. 369)

There must always be a parallel between a general theorem and a special example of a kind. In fact, I personally find — people are different; some people think abstractly very well — I don't. I always have to have examples to understand something the first time I hear it, and then I generalize from the examples. Other people like the general thing and then use it on the thing.

> – Esalen lecture, "Quantum Mechanical View of Reality (Part 1)," October 1984

I didn't do experiments, I never did, I just fiddled around. I made radios and gadgets. I fiddled around. Gradually, through books and manuals, I began to discover there were formulas applicable to, say, electricity in relating the current and resistance.

> – National Science Teachers Association Fourteenth Convention lecture, "What Is Science?" April 1966

The result of this is that I cannot remember anybody's name, and when people discuss physics with me, they often are exasperated when they say, "The Fitz-Gronin effect" and I ask, "what is the effect?" I can't remember the name.

– National Science Teachers Association Fourteenth Convention lecture, "What Is Science?" April 1966

I love the subject of physics and it has been my desire to try to share the delights of understanding it with any minds that were able to, male or female, and I don't believe that there's any reason, I never have believed that there's any reason that we know of, that there's a difference in the ability of one or another person to understand the physics.

– Oersted Medal acceptance speech, 1972

I'm sorry that I am unable to give you a recommendation for a nomination for the Niels Bohr Medal, but I have always made it a principle of never recommending or criticizing a colleague.

– Letter to Mr. Bjorn Andersen, February 1976

One day I'll be convinced there's a certain type of symmetry that everybody believes in, the next day I'll try to figure out the consequences if it's not, and everybody's crazy but me.

– *Omni* interview, February 1979

I don't know the effect I'm having. Maybe it's just my character; I don't know. I'm not a psychologist or sociologist, I don't know how to understand people, including myself.

– *Omni* interview, February 1979

On the program it says this is a keynote speech. I telephoned when I heard that I was going to give a keynote speech to ask, "What does that mean?" and they said that would be the after-dinner speech, and I said "No." So they moved to now, but they didn't change the title, and I don't know what a keynote speech is. I do not intend in any way to suggest what should be in this meeting as a keynote of the subjects or anything like that. I have my own things to say and to talk about, and there's no implication that anybody needs to talk about the same thing or anything like that.

– MIT conference, May 1981

All these things I don't understand — deep questions, profound questions; however, physicists have kind of a dopey way of avoiding all of these things.

– MIT conference, May 1981

Some new kind of thinking is necessary, but physicists, being kind of dull-minded, only look at nature and don't know how to think these new ways.

– MIT conference, May 1981

I never think, "This is what I like, and this is what I don't like," I think "This is what it is, and this is what it isn't," and whether or not I like it is really irrelevant, and I have extracted it out of my mind.

– "QED: Photons — Corpuscles of Light," Sir Douglas Robb Lectures, University of Auckland, June 1979

I will not be able to speak the way I usually speak because I speak too fast, so I will have to speak slowly and have not got time to say a great deal.

– "The Computing Machines in the Future," Nishina Memorial Lecture, August 1985

You tell them about how some people used to believe in witches, and of course nobody believes in witches now, and you say, "How could people believe in witches?" and then you turn around and you say, "Oh, what witches do we believe in now? What ceremonies do we do? Every morning we brush our teeth! What is the evidence that brushing teeth does us any good with cavities? As the Earth turns in orbit, there's an edge between light and dark, and on that edge, all the people — they're on that edge doing the same ritual for no good reason. Just like in the Middle Ages they had other rituals, and you're trying to picture this perpetual line of tooth brushers going around the Earth!

– Interview with Yorkshire Television program, "Take the World from Another Point of View," 1972

The pleasure in physics, for me, is that it's revealed that the truth is so remarkable and amazing. I have this disease, and many other people who have studied far enough to begin to understand a little about how things work are fascinated by it, and this fascination guides them on to such an extent that they've been able to convince governments to keep supporting them in this investigation that the race is making!

> – Interview with Yorkshire Television program, "Take the World from Another Point of View," 1972

I have tried many different ways of understanding the physical universe.

> – Letter to Richard D. Farley, August 1975

[Response to a colleague's suggestion for an autobiography:] Actually your suggestion has put a bee in my bonnet which I hope dies before it puts me to work.

> – Letter to Dr. Erik M. Pell, March 1976

It is my greatest pleasure to think anew about things, and I am delighted to discover that I have infected you with the same pleasure.

> – Letter to Dr. Frank Potter, November 1984 (*Perfectly Reasonable Deviations from the Beaten Track*, p. 371)

You ask me on what I think about life, etc., as if I had some wisdom. Maybe, by accident, I do — of course I don't know — all I know is I have opinions.

> – Letter to Mr. V. A. Van Der Hyde, July 1986 (*Perfectly Reasonable Deviations from the Beaten Track*, p. 413)

Thank you for your letter about my KNXT interview. You are quite right that I am very ignorant about smog and many other things, including the use of the finest English.

– Letter to Raymond Rogers, January 1966 (*Perfectly Reasonable Deviations from the Beaten Track*, p. 209)

There's a great deal of intimidation by intellectuals in a country of less intellectual people. It comes in the form of pompous studies and pompous words to describe ideas that are fairly simple or have very little content. If someone says they do not understand one of these ideas, they're often put down, which must be hard for those who don't have too much confidence in their own intelligence.

– *U.S. News and World Report* interview, February 1985

I can write under pressure. It's the only way I really can write.

– "Joy of the Chase," *The Daily Telegraph*, July 5, 1988

What about me? I had to come right back here to teach classes after being waited on hand and foot by two beautiful nymphs and their assistant commandos. It was tough.

– Letter to Mariela Johansen, January 1975

My wife and I think I am crazy.

– Letter to Mariela Johansen, January 1975

If you go in the same direction as everybody else, you have a whole host of people that you gotta get ahead of.

– CERN talk, December 1965

Those people who have for years been insisting in the face of all obvious evidence to the contrary that the male and female are equally capable of rational thought may have something.

– National Science Teachers Association Fourteenth Convention lecture, "What Is Science?" April 1966

I decided to stay at Caltech forever. I couldn't bear missing guys running around all excited with discoveries.

– *The Daily Times*, October 5, 1966

I know that that fraternity was a very important thing in my life. I know it — I mean, as far as social things — because, although it was hard to do, it forced me to do it. It's easy to not do it, it's scary, and it's easy to not to do it, but they made sure I did it. They taught me to dance. And so the confidence came relatively rapidly after a while.

– Interview with Charles Weiner, March 5, 1966 (Niels Bohr Library and Archives with the Center for the History of Physics)

[On English classes:] Yes. I didn't see why I had to worry about that now. After all, how do you spell something? Suppose I make a mistake? (This was my attitude — it was the attitude at the time.) You make a mistake in spelling. What does it mean? It means that the damn language is irrational. It's just a stupid method of spelling. Some guy ought to make some progress. If those English professors would sit around and figure out how to straighten out the spelling, instead of teaching this idiocy all the time — they had no feeling of progress, no feeling of development, like the sciences or anything.

– Interview with Charles Weiner, March 5, 1966 (Niels Bohr Library and Archives with the Center for the History of Physics)

[On college courses:] I'm trying to emphasize that my humanities effort was always somehow to figure out how I could, by using science, escape the humanities. I fought it to the bitter end.

  – Interview with Charles Weiner, March 5, 1966 (Niels Bohr Library and Archives with the Center for the History of Physics)

I cannot conceive that antagonism could result from the way I expressed myself, but only perhaps from the fact that I did express myself.

  – Letter to Bill Whitley (KNXT), May 1959 (*Perfectly Reasonable Deviations from the Beaten Track*, p. 101)

The physical world was real, and the mathematics, I had become enthralled with, but not for itself, really — you know what I mean? It was fascinating, but my real heart was somewhere else. So I decided, I have to get my hands dirty, I can't stand these abstract things. So I changed to electrical engineering, because there was something that was real. But then some few months later, I realized I'd gone too far, and that somewhere in between — that physics was the right place. So I moved around a little bit at the beginning, and ended up with the physics course.

  – Interview with Charles Weiner, March 4, 1966 (Niels Bohr Library and Archives with the Center for the History of Physics)

I went to a scientific school, MIT. And the fraternity, when you first join, they try to, if you think you're smart, keep you from feeling that you're too smart by giving you what looked like simple questions, to try to figure out what actually happens. It's like training for imagination. It's kind of fun.

– BBC, "Fun to Imagine" television series, 1983

I always tried to do it myself, because I'd learn something, maybe get a different idea. I never looked it up.

– Interview with Charles Weiner, March 5, 1966 (Niels Bohr Library and Archives with the Center for the History of Physics)

It turned out that the notes I took at conferences were never very useful for anything, and I don't take much notes at conferences any more.

– Interview with Charles Weiner, February 4, 1973 (Niels Bohr Library and Archives with the Center for the History of Physics)

You've got plenty of room and not too many people, and it looks like it ought to be good! Anyway, you do, and don't forget you have Rutherford, so it's okay!

– On life in New Zealand, "QED: Photons — Corpuscles of Light," Sir Douglas Robb Lectures, University of Auckland, June 1979

Yes, I shall write an article on "Energy" for you. However, instead of the proposed compensation of $225 plus 25 copies of my contribution, I should instead like to receive one set of the complete new encyclopaedia when it comes out.

– Letter to Warren E. Preece (*Encyclopaedia Britannica*), January 1970

People often think I'm a faker, but I'm usually honest, in a certain way — in such a way that often nobody believes me!

– *Surely You're Joking, Mr. Feynman!*, p. 41

My talents do not include competence for advising in International Relations.

– Letter to Clarence Streit, January 1966

Any vague theory that is not completely absurd can be patched up by more vague talk at every point that brings up inconsistencies — and if we begin to believe in the talk rather than in the evidence we will be in a sorry state.

– *Feynman Lectures on Gravitation*, p. 22

So I graduated. And I had to wear an academic outfit in order to graduate. And also I remember that they teased me, that Princeton didn't know what they were going to get — that Princeton was an elegant place, and I was just, you know, a rough guy, and so on. Not really worry about it, but I did take it seriously, that there was a matter of — you know, Princeton has a certain elegance. And I was not an elegant person.

– Interview with Charles Weiner, March 5, 1966 (Niels Bohr Library and Archives with the Center for the History of Physics)

I was OK, but I was kind of a rough, kind of a simple character, as far as society goes. But I wasn't worried about it. I was just sort of half-proud of it.

– Interview with Charles Weiner, March 5, 1966 (Niels Bohr Library and Archives with the Center for the History of Physics)

It used to worry me — how do I look, what am I like? So I got out of that when I was at MIT. Probably the usual growing up, although I have reasons, I claim to understand it a little bit in terms of this fight and so on. But anyway I did change at MIT, my personality, my fear of girls; my young and timid frightened and somewhat insecure character disappeared.

> – Interview with Charles Weiner, March 5, 1966 (Niels Bohr Library and Archives with the Center for the History of Physics)

I believe, although I can't guarantee it, that most of my education as a graduate student was through my own studying, through worrying about problems, through talking to friends, and very little courses. And that's the way it was in those days.

> – Interview with Charles Weiner, March 5, 1966 (Niels Bohr Library and Archives with the Center for the History of Physics)

When I was at MIT, I had read a lot. I didn't explain that. I forgot to. I was in the library a lot. I read advanced books. That's the way I taught myself, I read lots of stuff — I was very avid for reading and studying and learning. I read about general relativity, I learned it from a book, and I read a lot of quantum mechanics along with Welpin, and all this stuff, by reading.

> – Interview with Charles Weiner, March 5, 1966 (Niels Bohr Library and Archives with the Center for the History of Physics)

I had a great faith in that way of looking at the world — scientific — make sure what the reality is, don't get mixed up.

> – Interview with Charles Weiner, March 5, 1966 (Niels Bohr Library and Archives with the Center for the History of Physics)

I can tell an amusing story — you can always later throw it away, you know? I finally arrived in Ithaca at 2 a.m. or 12 something, in the night. I got off the train, and I slung my suitcase onto my shoulder as I always used to. Then I said: "Wait a minute now. You're a professor, and you have to try and behave like one." A porter asked me: "Can I carry your suitcase?" "No, I carry my own." Then I realized: I've got to start living in a dignified way. So I let him carry it to a taxi, and I sat rather elegantly in the back of the taxi, and the guy says, "Where to?" I say, "Biggest hotel in town, please."

– Interview with Charles Weiner, June 27, 1966 (Niels Bohr Library and Archives with the Center for the History of Physics)

You see, I'm a one-sided fellow. I understand and love the sciences. But there are many fields of intellectual things that I don't really go for, like literature, psychology, philosophy, and so on, unless it's done in a very sort of scientific way. I'm very one-sided. I'm not a wide guy, only very wide in the sciences, but very much in the sciences and limited.

– Interview with Charles Weiner, June 27, 1966 (Niels Bohr Library and Archives with the Center for the History of Physics)

I don't like the non-technical problems. They're not to my liking. I just don't feel good about them. So I don't have much to do with it and have never, therefore, done much consulting.

– Interview with Charles Weiner, June 28, 1966 (Niels Bohr Library and Archives with the Center for the History of Physics)

I don't mean I don't believe in thinking about what I'm doing.
Talking about it is another matter. I don't like to talk about it
anymore.

– Interview with Charles Weiner, June 28, 1966 (Niels Bohr
  Library and Archives with the Center for the History of
  Physics)

I was consulted on a problem of safety for nuclear energy
peacetime plants. I was thinking of designing one at the General
Electric Co. The reason that I was involved, of course, was
because I was involved in the safety at Los Alamos. And so I felt,
you ought to help out if you know about these things for peacetime
use in business.

– Interview with Charles Weiner, June 28, 1966 (Niels Bohr
  Library and Archives with the Center for the History of Physics)

I'm a very complicated man. I mean, I've got all kinds of side
things, infinite amounts of them.

– Interview with Charles Weiner, June 27, 1966 (Niels Bohr
  Library and Archives with the Center for the History of Physics)

Bob Wilson came in and said that he had been funded to do a job
that was a secret and he wasn't supposed to tell anybody, but he
was going to tell me because he knew that as soon as I knew what
he was going to do, I'd see that I had to go along with it.

– "The Remarkable Dr. Feynman," *Los Angeles Times Magazine*,
  April 20, 1986

But what I did immorally, I would say, was not to remember the reason that I said I was doing it, so when the reason changed — Germany was defeated — not the singlest thought came to my mind about that that now I have to reconsider why I am continuing to do this. I simply didn't think, okay?

– BBC, "The Pleasure of Finding Things Out," 1981

Theoretical physics is a human endeavor, one of the higher developments of human beings — and this perpetual desire to prove that people who do it are human by showing that they do other things that a few other humans do (like playing bongo drums) is insulting to me.

– Letter to Tord Pramberg, January 1967, regarding photos in a physics book of Feynman drumming (*Perfectly Reasonable Deviations from the Beaten Track*, p. 230)

You know, the real problem — you have to appreciate, really, all my time in these schools and around there — was the lack of supplies. Perhaps it was good, so that I had plenty of time to worry about elementary things before I was swamped by advanced things. I couldn't get books. The library had no calculus book. When they got it, it was within a week or two, I'm sure, that I took it out — the first guy to take it out. The first calculus book that was in town.

– Interview with Charles Weiner, March 4, 1966 (Niels Bohr Library and Archives with the Center for the History of Physics)

That was also what you worried about — that girls would think you were a sissy. Dumb, but that's life.

– Interview with Charles Weiner, March 5, 1966 (Niels Bohr Library and Archives with the Center for the History of Physics)

I am not happy as a member of a self-perpetuating honorary society.

– Letter to Dr. Detlev W. Bronk and the National Academy of Sciences, August 1961 (*Perfectly Reasonable Deviations from the Beaten Track*, p. 108)

At the very beginning, we had terribly important secrets. We'd worked out lots of stuff about uranium, how it worked, and all this stuff was in documents that were in filing cabinets that were made out of wood that had on them little ordinary, common padlocks.

– UCSB talk, "Los Alamos from Below," February 1975

I do remember one moment. I was in my pajamas, working on the floor with papers all around me, these funny-looking diagrams of blobs with lines sticking out. I said to myself, wouldn't it be funny if these diagrams really are useful, and other people start using them, and *Physical Review* has to print these silly pictures?

– *Omni* interview, February 1979, regarding the Feynman Diagrams

It was radioactive; it was plutonium. And we stood at the door of this room talking about it. There was a new element that was made by man that had never existed on the earth before, except for a very short period possibly, at the very beginning.

– UCSB talk, "Los Alamos from Below," February 1975

I must explain that because I am a scientist does not mean that I have not had contact with human beings. Ordinary human beings. I know what they are like. I like to go to Las Vegas and talk to the showgirls and the gamblers and so on. I have banged around a lot in my life, so I know about ordinary people.

– "The Unscientific Age," John Danz Lecture Series, 1963

[When asked if he would like to read his prepared obituary:] I have decided it is not a very good idea for a man to read it ahead of time; it takes the element of surprise out of it.

– "The Cult of Richard Feynman," *Los Angeles Times Magazine*, December 2, 2001, p. 16

Art, Music, and Poetry

For far more marvelous is the truth than any artist of the past imagined it. Why do the poets of the present not speak of it? What men are poets who can speak of Jupiter if he were a man, but if he is an immense spinning sphere of methane and ammonia must be silent?

– *Feynman Lectures on Physics*, vol. 1, pp. 3–6

I understood at last what art is really for, at least in certain respects. It gives somebody, individually, pleasure. You can make something that someone likes so much that they're depressed, or they're happy, on account of that damn thing you made! In science, it's sort of general and large: You don't know the individuals who have appreciated it directly.

– *Surely You're Joking, Mr. Feynman!*, p. 267

If you mean that this is an age of science in the sense that in art, in literature, and in people's attitudes and understandings and so forth that science plays a large part, I don't think it is a scientific age at all.

– "The Unscientific Age," John Danz Lecture Series, 1963

I think most kids in high school are very mature. Because I know my friends who were in literature — you know, who wrote — and my friends who wrote plays, and they would read the great plays. They wouldn't read the children's plays, you know. It's the same thing. Anybody that's any good in high school already knows that they have to look at the real stuff.

– Interview with Charles Weiner, March 4, 1966 (Niels Bohr Library and Archives with the Center for the History of Physics)

I have a friend who's an artist and he's sometimes taken a view which I don't agree with very well. He'll hold up a flower and say, "Look how beautiful it is," and I'll agree. But then he'll say, "I as an artist can see how beautiful a flower is. But you, as a scientist take it all apart, and it becomes dull." I think he's kind of nutty. First of all, the beauty that he sees is available to other people — and to me too, I believe. Although I may not be quite as refined aesthetically as he is, I can appreciate the beauty of a flower. But at the same time, I see much more in the flower than he sees. I can imagine the cells inside, which also have a beauty. There's beauty not just at the dimension of one centimeter; there's also beauty at a smaller dimension. There are the complicated actions of the cells, and other processes. The fact that the colors in the flower have evolved in order to attract insects to pollinate is interesting; that means insects can see the colors. That adds a question: Does this aesthetic sense we have also exist in lower forms of life? There are all kinds of interesting questions that come from a knowledge of science, which only adds to the excitement and mystery and awe of the flower. It only adds. I don't understand how it subtracts.

– *What Do You Care What Other People Think?*, p. 11

I wanted very much to learn to draw, for a reason that I kept to myself: I wanted to convey an emotion I have about the beauty of the world.

– *Surely You're Joking, Mr. Feynman!*, p. 261

The artists of the Renaissance said that man's main concern should be for man, and yet there are other things of interest in the world. Even the artists appreciate sunsets, and the ocean waves, and the march of the stars across the heavens. There is then some reason to talk of other things sometimes.

– *The Character of Physical Law*, p. 13

Bongo playing has never been "music" to me — I don't read notes or know anything about conventional music — it has just been fun making noise to rhythm — not much "intelligence" in the intellectual sense is involved.

 – Letter to Dr. William L. McConnell, March 1975 (*Perfectly Reasonable Deviations from the Beaten Track*, p. 281)

I am sorry not to want to send you a drawing, because it has been my policy not to sell them to people who want them because it is a physicist who made them.

 – Letter to Dr. William L. McConnell, March 1975 (*Perfectly Reasonable Deviations from the Beaten Track*, p. 282)

Mr. Auden's poem only confirms his lack of response to Nature's wonders for he himself says that he would like to know more clearly what we "want the knowledge for." We want it so we can love nature more.

 – Letter to Mrs. Robert Weiner, October 1967 (*Perfectly Reasonable Deviations from the Beaten Track*, p. 248)

And the crassness of our time, so much lamented, is a crassness that can be alleviated only by art, and surely not by science without art. Art and poetry can remind the mind of beauty and gradually make life more beautiful.

 – Letter to Mrs. Robert Weiner, October 1967 (*Perfectly Reasonable Deviations from the Beaten Track*, p. 248)

I had never thought of that idea that many possible viewpoints not obviously equivalent is a sign also of greatness in art.

 – Letter to Jay A. Young, August 1966

The value of science remains unsung by singers, so you are reduced to hearing — not a song or poem, but an evening lecture about it. This is not yet a scientific age.

– "The Value of Science," December 1955

The fact that I beat a drum has nothing to do with the fact that I do theoretical physics.

– Letter to Tord Pramberg, January 1967, regarding photos in a physics book of Feynman drumming (*Perfectly Reasonable Deviations from the Beaten Track*, p. 230)

[On Nobel Prize work:] It is true that most of the work was done in the quiet and peaceful atmosphere at Telluride — a quiet and peace which I was perpetually trying to dispel by jungle rhythms.

– Letter to Erik M. Pell, November 1965

The human mind as it evolved, it evolved from an animal, and it evolves in a certain way that it is like any new tool in that it has its diseases and difficulties. It has its troubles, and one of the troubles is that it gets polluted by its own superstitions that it confuses itself, and that the discovery was finally made of a way to keep it sort of in line so that they can make a little progress in some direction rather than to go around in a circle and to force itself into a hole.

– Galileo Symposium, "What Is and What Should Be the Role of Scientific Culture in Modern Society," September 1964

Nature

Poets say science takes away from the beauty of the stars — mere globs of gas atoms. I too can see the stars on a desert night, and feel them. But do I see less or more? The vastness of the heavens stretches my imagination — stuck on this carousel my little eye can catch one-million-year-old light. A vast pattern — of which I am a part . . . . What is the pattern, or the meaning, or the why? It does not do harm to the mystery to know a little about it.

– *Feynman Lectures on Physics*, vol. 1, pp. 3–10

The point of my remarks about poets was not meant to be a complaint that modern poets show no interest in modern physics — but that they show no emotional appreciation for those aspects of Nature that have been revealed in the last four hundred years.

– Letter to Mrs. Robert Weiner, October 1967 (*Perfectly Reasonable Deviations from the Beaten Track*, p. 248)

Physicists are trying to find out how nature behaves.

– *Omni* interview, February 1979

If we look out from physics, which of course is the center of the universe, and see what surrounds us . . .

– Audio recording of Feynman Lectures on Physics, Lecture 3, October 3, 1961

If you want to burn a diamond in air, you can — but you're kind of dopey.

– Audio recording of Feynman Lectures on Physics, Lecture 1, September 26, 1961

But it illustrates that, if we screw up the geometry sufficiently, it is possible that all of gravitation is related in a way to pseudo forces.

– Audio recording of Feynman Lectures on Physics, Lecture 12, November 7, 1961

Things on a small scale behave nothing like a large scale at all. That's what makes physics hard, and very interesting.

– Audio recording of Feynman Lectures on Physics, Lecture 2, September 29, 1961

Nature, as a matter of fact, seems to be so designed that the most important things in the real world appear to be a kind of complicated accidental result of a lot of laws.

– *The Character of Physical Law*, p. 122

There is the value of the worldview created by science. There is the beauty and the wonder of the world that is discovered through the results of these new experiences.

– National Science Teachers Association Fourteenth Convention lecture, "What Is Science?" April 1966

The nucleus is very small. The atom in its diameter is 10^{-8} cm, and the nucleus in its diameter is in the order of magnitude is 10 to the $^{-13}$ cm. What does that mean? If you had an atom, and you wanted to see the nucleus, you'd have to expand it, magnify it, so the whole atom is the size of this room and then the nucleus is a bare speck that you can just about make out with the eye — 1/100th of an inch across.

– Audio recording of Feynman Lectures on Physics, Lecture 2, September 29, 1961

So close is life to life. The universality of the deep chemistry of living things is indeed a fantastic and beautiful thing. And all the time we human beings have been too proud to recognize our kinship with the animals.

– "The Uncertainty of Science," John Danz Lecture Series, 1963

It is easy to demonstrate that if Nature was nonrelativistic, then if things started out that way then it would be that way for all time, and so the problem would be pushed back to Creation itself, and God only knows how that was done.

– Dirac Memorial Lectures, "The Reason for Antiparticles," 1986

For a successful technology, reality must take precedence over public relations, for nature cannot be fooled.

– Report of the Presidential Commission on the Space Shuttle *Challenger* Accident, Volume 2: Appendix F, June 1986

There's a very large number of stars in the galaxy. If you tried to name them, one a second, naming all the stars in our galaxy — and I don't mean all the stars in the universe, just this galaxy here — it takes three thousand years. And yet, that's not a very big number. Because if those stars were to drop one dollar bill on the earth during a year, each star dropping one dollar bill, they might take care of the deficit which is suggested for the budget of the United States.

– BBC, "Fun to Imagine" television series, 1983

[In response to a child's question, "What happens when an irresistible force meets an immovable object?":] They shake hands.

– From notes for "About Time" program, 1957

All things that we make are Nature. We arrange it in a way to suit our purpose.

– "The Computing Machines in the Future," Nishina Memorial Lecture, August 1985

We drop an egg on the sidewalk: it splatters in all directions. On the other hand if we had a smear of egg on the sidewalk, we would not expect it to come together to form a complete egg and ride back into our hand. So it would be obvious that the laws of nature appear different if we were to reverse the direction of time.

– From notes for "About Time" program, 1957

The biological example of writing information on a small scale has inspired me to think of something that should be possible. Biology is not simply writing information; it is doing something about it. A biological system can be exceedingly small. Many of the cells are very tiny, but they are very active; they manufacture various substances; they walk around; they wiggle; and they do all kinds of marvelous things — all on a very small scale.

– "There's Plenty of Room at the Bottom," December 1959

The origin of the force of gravitation is a problem which puzzles me too, and I do not think that we thoroughly understand it at all.

– Correspondence with R. I. Elliott on gravitation, January 1949

Trying to understand the way nature works involves a most terrible test of human reasoning ability. It involves subtle trickery, beautiful tightropes of logic on which one has to walk in order not to make a mistake in predicting what will happen.

– "The Uncertainty of Science," John Danz Lecture Series, 1963

Imagination reaches out repeatedly trying to achieve some higher level of understanding, until suddenly I find myself momentarily alone before one new corner of nature's pattern of beauty and true majesty revealed. That was my reward.

– From Les Prix Nobel en 1965 [Nobel Foundation], Stockholm, 1966

For instance, I stand at the seashore, alone, and start to think.
There are the rushing waves
mountains of molecules
each stupidly minding its own business
trillions apart
yet forming white surf in unison.
Ages on ages

before any eyes could see
year after year
thunderously pounding on the shore as now.
For whom, for what?
On a dead planet, with no life to entertain.
Never at rest
tortured by energy
wasted prodigiously by the sun
poured into space.
A mite makes the sea roar.
Deep in the sea
all molecules repeat
the patterns of one another
till complex new ones are formed.
They make others like themselves
and a new dance starts.
Growing in size and complexity
living things
masses of atoms
DNA, protein
dancing a pattern ever more intricate.
Out of the cradle
onto the dry land
here it is
standing:
atoms with consciousness;
matter with curiosity.
Stands at the sea
wonders at wondering: I

a universe of atoms
an atom in the universe.

  – "The Value of Science," December 1955

While we're admiring the human mind, which is something we enjoy doing — it's always someone else's human mind that we enjoy admiring — we should take some time off to stand in awe of nature, of a nature which can follow with such completeness and generality such an elegantly simple principle as the law of gravitation.

  – Audio recording of Feynman Lectures on Physics, Lecture 7,
    October 17, 1961

It is characteristic of the physical laws that that have this abstract character. Just like the conservation of energy is a theorem about quantities which you have to add together without a machinery, so the great laws of mechanics are quantitative mathematical laws for which no machinery is available.

  – Audio recording of Feynman Lectures on Physics, Lecture 7,
    October 17, 1961

[On his father:] He went on to say that in the world whenever there is any source of something that could be eaten to make life go, some form of life finds a way to make use of that source; and that each little bit of left over stuff is eaten by something.

  – National Science Teachers Association Fourteenth Convention
    lecture, "What Is Science?" April 1966

We have learned from much experience that all philosophical intuitions about what nature is going to do fail.

– The Character of Physical Law, p. 53

We used to laugh at the Greeks who claimed that the planets had to go in circles because it was a perfect figure. If they were talking in the modern times they would use group theoretic arguments and would imply that from a point of view of the planet the sun looks always the same, or that we have invariance under a combined time displacement and rotation. But the planets do not go in circles! Nature is not "symmetrical" and the question is why not?

– Programme of American Physical Society Annual Meeting, 1950

Nature uses only the longest threads to weave her patterns, so each small piece of her fabric reveals the organization of the entire tapestry.

– The Character of Physical Law, p. 34

The apparent enormous complexities of nature, with all its funny laws and rules, each of which has been carefully explained to you, are really very closely interwoven. However, if you do not appreciate the mathematics, you cannot see, among the great variety of facts, that logic permits you to go from one to the other.

– The Character of Physical Law, p. 41

We look at the stars: all the light that we see, that little tiny and influent spreads from the star over this enormous distance of three light years, for the nearest star. On, on, on, this light from the star is spreading, the wave-fronts are getting wider and wider, weaker and weaker, weaker and weaker, out into all of space, and finally, the tiny fraction of it comes in one square eighth of an inch, a tiny little black hole, and does something to me so I know it's there.

– BBC, "Fun to Imagine" television series, 1983

All the intellectual arguments that you can make will not communicate to deaf ears what the experience of music really is.

– *The Character of Physical Law*, p. 58

What is it about nature that lets this happen, that it is possible to guess from one part what the rest is going to do? That is an unscientific question: I do not know how to answer it, and therefore I am going to give an unscientific answer. I think it is because nature has a simplicity and therefore a great beauty.

– *The Character of Physical Law*, p. 173

There is always another way to say the same thing that doesn't look at all like the way you said it before. I don't know what the reason for this is. I think it is somehow a representation of the simplicity of nature.

– From *Nobel Lectures, Physics 1963–1970*, Elsevier Publishing Company, Amsterdam, 1972

Perhaps a thing is simple if you can describe it fully in several different ways without immediately knowing that you are describing the same thing.

– From *Nobel Lectures, Physics 1963–1970*, Elsevier Publishing Company, Amsterdam, 1972

Nature has always looked like a horrible mess, but as we go along we see patterns and put theories together; a certain clarity comes and things get simpler.

– *QED: The Strange Theory of Light and Matter*, p. 149

You can know the name of a bird in all the languages of the world, but when you're finished, you'll know absolutely nothing whatever about the bird. You'll only know about humans in different places, and what they call the bird. So lets look at the bird and see what it's doing — that's what counts.

– *What Do You Care What Other People Think?*, p. 14

If you take an apple and magnify it to the size of the earth, then the atoms in the apple are approximately the size of the original apple.

– *Feynman Lectures on Physics*, Vol. 1, pp. 1–3

Nature doesn't care what we call it, she just keeps on doing it whatever way she wants.

– Audio recording of Feynman Lectures on Physics, Lecture 1, September 26, 1961

The fact that there's no perpetual motion at all is a general statement of the conservation of energy law.

   – Audio recording of Feynman Lectures on Physics, Lecture 4, October 6, 1961

If you want to talk about Nature, you're going to talk about something complicated and dirty, and therefore at first approximations, ever increasing in accuracy.

   – Audio recording of Feynman Lectures on Physics, Lecture 12, November 7, 1961

Nature is, no doubt, simpler than all our thoughts about it.

   – Yorkshire Television interview, "Take the World from Another Point of View," 1972

I'm trying to find out, not how nature could be, but how it is. See what's right.

   – Yorkshire Television interview, "Take the World from Another Point of View," 1972

And so it can be that distances and times all seem to change, and things will be frightening and disturbing and disguise a sense of time as different from your sense of time, but its only because there is something left. There's another view in which the time is added together with the space, which makes this a new sense but requires an enormous amount of imagination, because we haven't any analogous experience of this kind.

   – Audio recording of lecture on relativity, Douglas Advanced Research Laboratory, 1967

If it were a common experience for biological systems to move that fast, then no doubt inside the brain there would be evolved a special wiring that would not, in fact, require us to learn the theory of relativity; it would be an innate feeling that this is right.

– Audio recording of lecture on relativity, Douglas Advanced Research Laboratory, 1967

This rather trivial understanding of relativity could have easily been understood the moment perspective was understood, and the famous old kingdom legends about the guys who feel the elephants, you know, as a rope, because he's holding on to the tail, or as a leaf because he's holding on to the ear, and so on, is just simply the same idea that things depend upon your point of view.

– Audio recording of lecture on relativity, Douglas Advanced Research Laboratory, 1967

Energy is a very subtle concept. It is very, very difficult to get right.

– National Science Teachers Association Fourteenth Convention lecture, "What Is Science?" April 1966

Is it unbelievable that a part of the brain can make do much "thinking" and interpreting and we would be unaware of it and not be able to control it? Maybe. But maybe not. The simplest animal must be able to think at this kind of level if he sees at all.

– On human and animal eyesight, in a letter to Edwin H. Land (Polaroid Corporation), May 1966 (*Perfectly Reasonable Deviations from the Beaten Track*, p. 224)

Relativity is not simply that things depend upon your point of view; it is.

– Audio recording of lecture on relativity, Douglas Advanced Research Laboratory, 1967

Light has weight, a finite weight proportional to its finite mass — neither zero nor infinity. Any object as it speeds up gets heavier until at light speed it is infinitely heavier than it would be at rest. But light never is at rest so the argument doesn't apply to light. In gravitational fields light falls, very slightly, and the images of stars seen very near the sun (during an eclipse) are slightly out of place because the light doesn't go in straight lines, but curves slightly down toward the sun as it goes by, because it falls.

– From notes for "About Time" program, 1957

The world is a spinning ball, and people are held on it on all sides, some of them upside down. And we turn like a spit in front of a great fire. We whirl around the sun. That is more romantic, more exciting. And what holds us? The force of gravitation, which is not only a thing of the earth, but is the thing that makes the earth round in the first place, holds the sun together, and keeps us running around the sun in our perpetual attempt to stay away. This gravity holds its sway not only on the stars, but between the stars; it holds them in the great galaxies for miles and miles in all directions.

– "The Uncertainty of Science," John Danz Lecture Series, 1963

The principle of relativity can be stated this way: that the motion of bodies among themselves in an enclosed space is the same whether the enclosed space is standing still or is moving uniformly at a constant speed in a straight line.

– Audio recording of lecture on relativity, Douglas Advanced Research Laboratory, 1967

The internal machinery of life, the chemistry of the parts, is something beautiful.

– "The Uncertainty of Science," John Danz Lecture Series, 1963

What looks still to our crude eyes is a wild and dynamic dance.

– "The Uncertainty of Science," John Danz Lecture Series, 1963

All other aspects and characteristics of science can be understood directly when we understand that observation of science is the ultimate and final judge of the truth of an idea.

– "The Uncertainty of Science," John Danz Lecture Series, 1963

If you have any appreciation for the complexities of nature and for the evolution of life on earth, you can understand the tremendous variety of possible forms that life would have.

– "The Unscientific Age," John Danz Lecture Series, 1963

Our psychological feeling of the flow of time then has been converted to a definite physical idea of a quantity which we can measure accurately and can speak of equal intervals of time.

– From notes for "About Time" program, 1957

So now when we ask how old is the universe we had better ask where.

– From notes for "About Time" program, 1957

The question is whether two events which appear to occur simultaneously to me, appear also simultaneously to anyone else. This question is the question as to whether there is an absolute meaning to "the present" because the present is all those events which are occurring at the same time as "now." If a different set of events occur "now" from your point of view then we will have to say that the present for me and you is not the same and has no absolute significance.

– From notes for "About Time" program, 1957

You are lucky you have such a deep interest in nature, and even if you find it is much more complicated and difficult to understand than you thought, when you learn more about it, it is also in certain ways more simple and beautiful than what you can imagine.

– Letter to student Charles E. Tucker, April 1967

You all know something about the wonders of science — it isn't a popular audience I'm talking to — so I won't try to make you enthusiastic once again with the facts about the world; the fact that we are all made of atoms, the enormous ranges of time and space that there are, the position of ourselves historically is a result of a remarkable series of evolution, the position of ourselves in the evolutionary sequence, and further the most remarkable aspect of our scientific world view is its universality in this sense, although we talk about our being specialists that we are really not.

– Galileo Symposium, "What Is and What Should Be the Role of Scientific Culture in Modern Society," September 1964

That the fact that our knowledge is in fact universal is something that is not completely appreciated, that the position of the theories are so complete that we hunt for exceptions and we find them very hard to find — in the physics at least — and the great expense of all these machines and so on is to find some exception to what is already known.

– Galileo Symposium, "What Is and What Should Be the Role of Scientific Culture in Modern Society," September 1964

Do you have any ideas about the mesons and fundamental particles? The meson theory ideas I have are either just phenomenological observations or childish modifications of electrodynamics. Surely Nature has a better imagination than to use the pseudoscalar field theory!

– Letter to Professor L. Landau, November 1954

There was another thing someone could measure in classical physics, which was something like the speed — actually it's called the momentum: it's the speed times the mass, which tells you how well something coasts!

– Esalen lecture, "Quantum Mechanical View of Reality (Part 1)," October 1984

So there came a time, perhaps, when for some species the rate at which learning was increased reached such a pitch that suddenly a completely new thing happened; things could be learned by one animal, passed on to another and another fast enough that it was not lost to the race. Thus became possible an accumulation of knowledge of the race. This has sometimes been called time-binding.

– National Science Teachers Association Fourteenth Convention
lecture, "What Is Science?" April 1966

This phenomenon of having a memory for the race, of having an accumulated knowledge passable from one generation to another, was new in the world. But it had a disease in it. It was possible to pass on mistaken ideas. It was possible to pass on ideas which were not possible to the human race.

– National Science Teachers Association Fourteenth Convention
lecture, "What Is Science?" April 1966

Then a way of avoiding the disease was discovered. This is to doubt that what is being passed from the past is in fact true, and to try to find out *ab initio*, again from experience, what the situation is, rather than trusting the experience of the past in the form in which it was passed down.

– National Science Teachers Association Fourteenth Convention
lecture, "What Is Science?" April 1966

The trees are made of air, primarily. When they are burned, they go back to air, and in the flaming heat is released the flaming heat of the sun that was bound in to convert the air into tree, and in the ash is the small remnant of the part which did not come from air, that came from the solid earth.

– National Science Teachers Association Fourteenth Convention lecture, "What Is Science?" April 1966

And I'm not happy with all the analyses that go with just the classical theory because nature isn't classical, dammit, and if you want to make a simulation of nature, you'd better make it quantum mechanics, and by golly it's a wonderful problem because it doesn't look so easy.

– MIT conference, May 1981; "Simulating Physics with Computers"

There's an enormous range of this one property — the wavelength — a range of phenomena that's this complete, enormous spectrum. The eye sees a very narrow range of this spectrum, and it's all put together with this one theory of electromagnetic waves. I'm going to talk about that part of it, and I'm going to call it "light" instead of saying "electromagnetic radiation." Light, what we see, is only one little part, but the physicist's point of view is the accident that the human eye happens to be sensitive to waves from here to here is not essential. The phenomena are the same — it's a whole range!

– "QED: Photons — Corpuscles of Light," Sir Douglas Robb Lectures, University of Auckland, June 1979

If you want to know the way nature works, we looked at it happily, and that's the way it looks! You don't like it? Go somewhere else — to another universe where the rules are simpler!

– "QED: Photons — Corpuscles of Light," Sir Douglas Robb Lectures, University of Auckland, June 1979

It's always possible that tomorrow somebody's going to figure it out, but it's going to be very difficult and very strange.

– "QED: Photons — Corpuscles of Light," Sir Douglas Robb Lectures, University of Auckland, June 1979

As much as we admire the human mind, we must also stand in awe of Nature which follows with such completeness and generality such an elegantly simple principle as this law of gravitation. And what is this law? That every object in the universe attracts every other with a force that is proportional to the mass of each and varies inversely as the square of the distance between them.

– In personal notes

Things that are very common and observed all the time and which appear perfectly obvious are quite different in this world. It turns out that what we thought was obvious is wrong, and it's much more complicated — or not more complicated but just different! In fact, sometimes it's simpler and more beautiful.

– "QED: Electrons and Their Interactions," Sir Douglas Robb Lectures, University of Auckland, June 1979

Most elastic things, like steel springs and so on, is nothing but this electrical thing pulling back: You pull the atoms a little bit apart when you bend something, and then they try to come back together again. But rubber bands work on a different principle. There are some long molecules, like chains, and other little ones that are shaking all the time, that are bombarding them. And the chains are all kind of kinky. When you pull open the rubber band, the strings get straighter. But these strings are being bombarded on the side by other atoms that are trying to shorten them by kinking them. So it pulls back, it's trying to pull back. It pulls back only because of the heat!

– BBC, "Fun to Imagine" television series, 1983

I've always found rubber bands fascinating. To think, when they've been sitting on an old package of papers for a long time, holding those papers together, its done by a perpetual pounding-pounding-pounding of the atoms against these chains, trying to kink them and trying to kink them, year after year.

– BBC, "Fun to Imagine" television series, 1983

The world is a dynamic mess of jiggling things.

– BBC, "Fun to Imagine" television series, 1983

We're so used to circumstances in which these electrical phenomena are all canceled out, everything is sort of neutral, where pushing and pulling is sort of dull, but nature has these wonderful things: magnetic forces and electrical forces.

– BBC, "Fun to Imagine" television series, 1983

You must compare your ideas with Nature; she tells you yes or no. She produces phenomena that require explanation. You cannot make your own assumptions and analyze the consequences.

– Letter to Mr. Robert Bonic, January 1974

[On the laws of gravitation:] The only applications of the knowledge of the law that I can think of are in geophysical prospecting, in predicting the tides, and nowadays, more modernly, in working out the motions of the satellites and planet probes that we send up, and so on; and finally, also modernly, to calculate the predictions of the planets' positions, which have great utility for astrologists who publish their predictions in horoscopes in the magazines. It is a strange world we live in — that all the new advances in understanding are used only to continue the nonsense which has existed for 2,000 years.

– *The Character of Physical Law*, p. 27

Astronomy is older than physics. In fact, it got physics started by showing the beautiful simplicity of the motion of the stars and planets, the understanding of which was the beginning of physics.

– *Feynman Lectures on Physics*, vol. 1, p. 59

I think Nature's imagination is so much greater than man's, she's never gonna let us relax!

– BBC, "Fun to Imagine" television series, 1983

The most remarkable discovery in all of astronomy is that the stars are made of atoms of the same kind as those on Earth.

– *Feynman Lectures on Physics*, vol. 1, p. 3

Of course, men want knowledge for many other purposes also, to make war, to make a commercial success, to help the sick or the poor, etc., motives of various values. These obvious motives and their consequences the poets do understand and do write about. But emotions of awe, wonder, delight and love which are evoked upon learning Nature's ways in the animate and inanimate word, together (for they are one) is rarely expressed in modern poetry where the aspect of Nature being appreciate is one which could have been known to men in the Renaissance.

– Letter to Mrs. Robert Weiner, October 1967 (*Perfectly Reasonable Deviations from the Beaten Track*, p. 248)

How is it possible, by looking at a piece of nature, to guess how another part must look, where you've never been before? It's only in modern times that man has really been able to guess what nature is going to do in situations that he's never looked at before.

– BBC, "Strangeness Minus Three," 1964

This universe has been described by many, but it just goes on, with its edge as unknown as the bottom of the bottomless sea of the ancients' idea — just as mysterious, just as awe-inspiring, and just as incomplete as the poetic pictures that came before.

– "The Uncertainty of Science," John Danz Lecture Series, 1963

I like science because when you think of something you can check it by experiment: "Yes" or "No," nature says, and you go on from there progressively. Other wisdom has no equally certain way of separating truth from falsehood.

– Letter to Beata Kamp, February 1983 (*Perfectly Reasonable Deviations from the Beaten Track*, p. 356)

We are so used to looking at the world from the point of view of living things that we cannot understand what it means not to be alive, and yet most of the time the world had nothing alive on it. And in most places in the universe today there probably is nothing alive.

– "The Uncertainty of Science," John Danz Lecture Series, 1963

What makes you so sure that the new discovery of the interrelationship between nuclear forces is going to be so wonderful? How do we know it isn't going to be some complicated, dirty, or simple thing? We don't know. We keep on trying anyway. We're not sure. It's worth the risk. Because it very likely will be peculiar and if it is, it'll be very interesting.

– BBC, "Strangeness Minus Three," 1964

Imagination

I don't know why it is that some people find science dull and difficult, and other people find it fun and easy, but there's one characteristic that I get a big kick from, and that is that it takes so much imagination to try to figure out what the world is really like.

– BBC, "Fun to Imagine" television series, 1983

I would like not to underestimate the value of the world-view which is the result of scientific effort. We have been led to imagine all sorts of things infinitely more marvelous than the imaginings of poets and dreamers of the past. It shows that the imagination of nature is far, far greater than the imagination of man.

– "The Value of Science," December 1955

It was thought in the Middle Ages that people simply make many observations, and the observations themselves suggest the laws. But it does not work that way. It takes much more imagination than that. So the next thing we have to talk about is where the new ideas come from. Actually, it does not make any difference, as long as they come.

– "The Uncertainty of Science," John Danz Lecture Series, 1963

It is surprising that people do not believe that there is imagination in science. It is a very interesting kind of imagination, unlike that of the artist. The great difficulty is in trying to imagine something that you have never seen, that is consistent in every detail with what has already been seen, and that is different from what has been thought of; furthermore, it must be definite and not a vague proposition. That is indeed difficult.

– "The Uncertainty of Science," John Danz Lecture Series, 1963

The game I play is a very interesting one. It's imagination, in a tight straightjacket, which is this: that is has to agree with the known laws of physics.

– "Tiny Machines," Esalen Institute

I find myself trying to imagine all kinds of things all the time, and I get a kick out of it, just like a runner gets a kick out of sweating. I get a kick out of thinking about these things. I can't stop!

– BBC, "Fun to Imagine" television series, 1983

I think with science, one of the things that makes it very difficult is that it takes a lot of imagination.

– BBC, "Fun to Imagine" television series, 1983

The great part of astronomy is the imagination that's been necessary to guess what kind of structures, what kinds of things, could be happening to produce the light and the effects of the light, and so on, of the stars we do see.

– BBC, "Fun to Imagine" television series, 1983

Many times in science, by using imagination, you've imagined something which could be, according to all the known knowledge of the law, and you don't know whether it is yet or not.

– BBC, "Fun to Imagine" television series, 1983

We have to imagine how things might look from another point of view. A point of view that perhaps we have never been able to take.

– Audio recording of lecture on relativity, Douglas Advanced Research Laboratory, 1967

It requires careful study of the detailed experiments, and lots of thought, to lift our imagination to the point to see this. Otherwise, we would be fine. If we were able to move, say, very rapidly, then we would be able to see all these things quite directly.

– Audio recording of lecture on relativity, Douglas Advanced Research Laboratory, 1967

It's much more interesting for me (unless I'm working on it) to leave a mystery a mystery, rather than to make believe I know an answer to it.

– BBC, "The Pleasure of Finding Things Out," 1981

Knowledge is of no real value if all you can tell me is what happened yesterday. It is necessary to tell what will happen tomorrow if you do something not only necessary, but fun. Only you must be willing to stick your neck out.

– "The Uncertainty of Science," John Danz Lecture Series, 1963

Our imagination is stretched to the utmost, not as in fiction, to imagine things which are not really there, but just to comprehend those things which are there.

– *The Character of Physical Law*, pp. 127–128

The stuff of fantasizing in looking at the world, imagining things, which really isn't fantasizing because you just try to imagine the way it really is, comes in handy sometimes.

– BBC, "Fun to Imagine" television series, 1983

I gotta stop somewhere, to leave you something to imagine!

– BBC, "Fun to Imagine" television series, 1983

Humor

I am Professor Feynman, in spite of this suit-coat.

– Galileo Symposium, "What Is and What Should Be the Role of Scientific Culture in Modern Society," September 1964

A poet, I think it is, once said,
The whole universe is in a glass of wine
I don't think we'll ever know in what sense he meant that —
for the poets don't write to be understood —
but it is true that if you look at a glass of wine closely enough
you will see the entire universe.
There are the things of physics:
the twisting liquid, the reflections in the glass,
and our imagination adds the atoms.
It evaporates, depending on the wind and weather.
The glass is a distillation of the earth's rocks,
and in its composition, as we've seen,
the secret of the universe's age,
and the evolution of the stars.
What strange array of chemicals are in the wine?
How did they come to be?
There are the ferments, the enzymes,
the substrates, and the products.
And there, in wine, was found the great generalization:
All life is fermentation.
Nor can you discover the chemistry of wine
without discovering, as did Pasteur,
the cause of much disease.
How vivid is the claret, pressing its existence
into the consciousness that watches it.
And if our small minds, for some convenience,
divides this glass of wine, this universe, into parts —
the physics, biology, geology, astronomy, psychology, and all —
remember that Nature doesn't know it.
So we should put it all back together,

and not forget, at last, what it's for.
Let it give us one final pleasure more:
drink it up and forget about it all!

  – *Feynman Lectures on Physics*, pp. 3–10

Although my mother didn't know anything about science, she had a great influence on me as well. In particular, she had a wonderful sense of humor, and I learned from her that the highest forms of understanding we can achieve are laughter and human compassion.

  – *What Do You Care What Other People Think?*, p. 19

My wife couldn't believe I'd actually accept an invitation to give a speech where I'd have to wear a tuxedo. I did change my mind a couple of times.

  – *Omni* interview, February 1979

I am glad to hear you think the show is O.K., but very surprised to learn that I have a "professional approach." What profession? T.V. personality? Well, all I know I learned from my director. Maybe you mean by "professional approach" that I act like a steel wastecan can!

  – Correspondence with Philip Daly (BBC), August 1964

Thank you for the invitation to attend the "Conference on Color, Flavor, and Unification" to be held on November 30. Since you give the year 1975 and offer to pay airfare, may I assume we are taking a trip back in time via the *Enterprise*? I am happy to accept in any case.

– Letter to Dr. Gordon Shaw, May 1979

[While drawing arrows:] The arrows are called amplitudes, so I may say amplitude rather than arrow. It's just a word, and we can have any word we want, like it says in Lewis Carroll!

– Esalen lecture, "Quantum Mechanical View of Reality (Part 2)," October 1984

There are 10^{11} stars in the galaxy. That used to be a huge number. But it's only a hundred billion. It's less than the national deficit! We used to call them astronomical numbers. Now we should call them economical numbers.

– *Feynman Lectures on Physics*

I think that it is much more likely that the reports of flying saucers are the results of the known irrational characteristics of terrestrial intelligence rather than of the unknown efforts of extra-terrestrial intelligence.

– *The Character of Physical Law*, pp. 165–166

But you can appreciate the difficulties that the chemists have and also appreciate why the names are so long. It's not because they want to be obstinate, it's because they have an extremely difficult problem, to describe this thing in words. Why they don't just draw the pictures all the time, I don't know.

– Audio recording of Feynman Lectures on Physics, Lecture 1, September 26, 1961

What's wrong with that? That's an excellent way to do things: to take a good flying guess at it first and then check it.

– Audio recording of Feynman Lectures on Physics, Lecture 8, October 20, 1961

Everybody who comes to a scientific lecture knows they are not going to understand it, but maybe the lecturer has a nice, colored tie to look at. Not in this case!

– *QED: The Strange Theory of Light and Matter*, p. 9

Never say you'll give a talk unless you know clearly what will you will talk about and roughly what you will say.

– Notes

For those who want some proof that physicists are human, the proof is in the idiocy of all the different units which they use for measuring energy.

– *The Character of Physical Law*, p. 75

If we look up at the sky: to many ancient people, it appeared like the surface of a dome in which there are points of light. And this idea that it could be a shell with points of light on it is not self-evidently crazy. It's only the result of a great deal of astronomical observation that permits the idea that that's not what we're looking at.

– Audio recording of lecture on relativity, Douglas Advanced Research Laboratory, 1967

One of the biggest and most important tools of theoretical physics is the wastebasket.

— Future for Science interview

It was a complete surprise to me when he said, "I wanted to tell you that you won the prize." "Won?" I said, "hot dog!" You see? So he says, "It's interesting to hear a serious scientist saying something like, hot dog." I said, "Listen, you call up any serious scientist and tell him he won $15,000, he'll say hot dog."

— Interview with Charles Weiner, June 28, 1966 (Niels Bohr Library and Archives with the Center for the History of Physics)

Hence, blow on soup if you want to cool it off.

— Audio recording of Feynman Lectures on Physics, Lecture 1, September 26, 1961

You're on the air, they say. Of course, it's got nothing to do with the air. You can have radio broadcasts without any air.

— Audio recording of Feynman Lectures on Physics, Lecture 2, September 29, 1961

All these particles are really chicken tracks in a cloud chamber.

— Audio recording of Feynman Lectures on Physics, Lecture 2, September 29, 1961, Q&A

If you can't see gravitation acting here, you have no soul!

— Audio recording of Feynman Lectures on Physics, Lecture 7, October 17, 1961

The Greeks got somewhat confused — they were helped by, of course, some very confusing Greeks.

– Audio recording of Feynman Lectures on Physics, Lecture 8, October 20, 1961

The question is, how would you answer her if you were the cop? Well, if you're really the cop, then no subtleties are involved; it's very simple. You say, "tell that to the judge."

– Audio recording of Feynman Lectures on Physics, Lecture 8, October 20, 1961

I am human enough to tell you to go to hell.

– Letter to Tord Pramberg, January 1967, regarding photos in a physics book of Feynman drumming (*Perfectly Reasonable Deviations from the Beaten Track*, p. 230)

I have been thinking about educating the people who do cosmology and 1) can write Chapter 25 with a manuscript to you April 1, 1969, 2) can talk into a tape recorder and you can make what you can out of the mess, 3) hope you quit bothering me, you bastard. In case of the first two, all best regards.

– Letter to Allan Sandage, February 1969

I always get the impression when I come east that the east is backward. If we had a room this hot in the west, we would air-condition it.

– "Current Algebras and Strong Interactions," 1967

[After a microphone malfunction:] Now I predict that the probability that I'll have a microphone of that kind next time is very low!

– "QED: Fits of Reflection and Transmission," Sir Douglas Robb Lectures, University of Auckland, 1979

It is always possible to follow the right lines after the events.

– Programme of American Physical Society Annual Meeting, 1950, regarding priests' reading sheeps' liver lines to predict the future

When you work hard, there are moments when you think, "At last, I've discovered that mathematics is inconsistent!" But pretty soon you discover the error, as I finally did.

– *Feynman's Tips on Physics*, p. 63

[Standing in front of three press microphones:] Why three microphones? It is ridiculous. Why not tap three lines off one microphone, if you need three.

– Press conference, April 23, 1963

I get the compliments, and all you get is some remark like "and there was no buffoon commentator asking you dumb questions," or the like. Little did most people (even some in the business) realize how it was actually done. They all believed the illusion that all I had to do was open my mouth and talk for an hour.

Like all true art, the artist disappears, and it looks natural and wonderful.

    – Letter to Christopher Sykes (BBC), March 1983

My schedule is such lately that I must refuse to get bogged down reading someone else's theory; it may turn out to be wonderful and there I'd be with something else to think about.

    – Letter to Francis Crick, March 1978 (*Perfectly Reasonable Deviations from the Beaten Track*, pp. 317–318)

Something I think theoretical physicists should be ashamed of: when you think of how much money is being put into these experiments and the big apparatus and so forth, and here we just sit around with a beautiful theory and mumble about it and can't calculate any numbers! We shouldn't get our salaries, I think. Or maybe they should be raised — we'd work faster!

    – "QED: New Queries," Sir Douglas Robb Lectures, University of Auckland, 1979

There are certain kinds of men in every field that I can talk to as well as I can talk to a good scientist.

    – Interview with Yorkshire Television program, "Take the World from Another Point of View," 1972

If you give more money to theoretical physics, it doesn't do any good if it just increases the number of guys following the comet head.

– CERN talk, December 1965

So while I was a professor, I could act very much like a student, even a freshman. I could be mistaken for a freshman in a perfectly legitimate way.

– Interview with Charles Weiner, June 27, 1966 (Niels Bohr Library and Archives with the Center for the History of Physics)

If the doctor says, "This guy's got bumps sticking out," you do something. But he says, "He's a little bit nuts," you're afraid to ask questions. Very amusing.

– Interview with Charles Weiner, June 28, 1966 (Niels Bohr Library and Archives with the Center for the History of Physics)

Well, unfortunately, you see I have always been unfashionable. How this parton* thing has been so successful that I have become fashionable. I have to find an unfashionable thing to do.

– Interview with Charles Weiner, February 4, 1973 (Niels Bohr Library and Archives with the Center for the History of Physics)

---

*The parton model was proposed by Feynman and James Bjorken in the 1960s to address new discoveries on hadron collision.

This conference was worse than a Rorschach test: There's a meaningless inkblot, and the others ask you what you think you see, but when you tell them, they start arguing with you!

– *Surely You're Joking, Mr. Feynman!*

[In response to a fan letter:] I am now unique — a physicist with a fan who has fallen in love with him from seeing him on TV. Thank you, oh fan! Now I have everything anyone could desire. I need no longer be jealous of movie stars. [Letter signed: Your fan-ee (or whatever you call it — the whole business is new to me), Richard P. Feynman]

– Letter to Ilene Ungerleider, August 1975 (*Perfectly Reasonable Deviations from the Beaten Track*, p. 286)

I just love the smell of rats because it's the spoor of exciting adventure.

– Letter to Gweneth and Michelle Feynman, February 1986 (*Perfectly Reasonable Deviations from the Beaten Track*, p. 402)

So I have invented another myth for myself: that I'm irresponsible. I'm actively irresponsible. I tell everybody I don't do anything. If anybody asks me to be on a committee to take care of admissions, "No, I'm irresponsible. I don't give a damn about the students." Of course I give a damn about the students, but I know that somebody else will do it.

– Interview with Yorkshire Television program, "Take the World from Another Point of View," 1972

It is gratifying to be honored by one's friends and neighbors. I have found, however, that winning awards always brings about some pitfalls and obstacles. I had to go to Sweden to accept one award and had to get up at 7 a.m. to accept another.

– Chamber of Commerce Outstanding Citizen Award acceptance speech

This is a tremendous idea — that you should do some careful experiments instead of deep philosophical arguments and find something out.

– Audio recording of Feynman Lectures on Physics, Lecture 7, October 17, 1961

I therefore don't have much to say. But I will talk a long time anyway.

– Programme of American Physical Society Annual Meeting, 1950

There are also of course in the world a number of phenomena that you cannot beat that are just the result of a general stupidity. And we all do stupid things, and we know some people do more than others, but there is no use in trying to check who does the most.

– "The Unscientific Age," John Danz Lecture Series, 1963

[In response to a fan letter:] One gets a nice feeling when reading letters from people who enjoyed something you did, and the glow doesn't usually wear off by lunchtime.

– Letter to J. S. Paxton, January 1982

For example, there is the absurdity of the young these days chanting things about purple people eaters and hound dogs, something that we cannot criticize at all if we belong to the old flat foot floogie and a floy floy or the music goes down and around. Sons of mothers who sang about "come Josephine, in my flying machine," which sounds just about as modern as "I'd like to get you on a slow boat to China." So in life, in gaiety, in emotion, in human pleasures and pursuits, and in literature and so on, there is no need to be scientific, there is no reason to be scientific. One must relax and enjoy life.

– "The Unscientific Age," John Danz Lecture Series, 1963

It is odd, but on the infrequent occasions when I have been called upon in a formal place to play the bongo drums, the introducer never seems to find it necessary to mention that I also do theoretical physics.

– *The Character of Physical Law*, p. 13

Love

It's necessary to fall in love with a theory, and like falling in love with a woman, it's only possible if one does not completely understand her.

– CERN talk, December 1965

Beautiful blonde or no beautiful blonde, I am still not married so I can still work at physics.

– Letter to Dr. Ted A. Welton, October 1948

If I'm sitting next to a swimming pool and somebody dives in and she's not too pretty, so I can think of something else, I think of the waves and things that have formed in the water.

– BBC, "Fun to Imagine" television series, 1983

There are other things in which scientific methods would be of some value; they are perfectly obvious but they get more and more difficult to discuss — such things as making decisions. I do not mean that it should be done scientifically, such as in the United States the Rand Company sits down and makes arithmetical calculations. That reminds me merely of my sophomore days at college in which in discussing women we discovered that by using electrical terminology — impedance, reluctance, resistance — that we had a deeper understanding of the situation.

– Galileo Symposium, "What Is and What Should Be the Role of Scientific Culture in Modern Society," September 1964

[On what he was most proud of:] I was able to love my first wife with as deep a love as I was able to.

– The *Los Angeles Times*, February 16, 1988

My mother taught me, said I must step out of the bus first and help her out, and all this stuff, and I worried about: what am I going to talk about? I still remember what we talked about. It's so silly, because, you know this first experience. She asked me if I played the piano, and I told her I had tried to learn, and I used to take lessons, for a little while. After I was older — after many long months of this I could only play something called "Dance of the Daisies," or fairies or something, and this didn't seem to me a very good thing, and so I didn't do piano. This and that, we talked about. Later, as we were saying good-bye, she said, "Thank you for a lovely evening." I was so impressed. I was so happy. Then I found out, on my second date, that the girl said, "Thank you for a lovely evening." On my third date, when we were saying good-night, just at the door, I said to her, "Thank you for a lovely evening," and she got paralyzed, unable to say anything, because that was what she was just about to say. So I quickly learned the formal from the truth, you see.

– Interview with Charles Weiner, March 4, 1966 (Niels Bohr
  Library and Archives with the Center for the History of Physics)

I was frightened of girls when I went there. I remember when I had to deliver the mail. I'm just trying to tell you the differences. It's interesting, how social attitudes develop. When I had to deliver the mail; to take the mail out from upstairs. It happened to be a time when some of the juniors had a few girlfriends, two girlfriends. There were just sitting on the steps talking, and I just didn't know how the heck I was going to be able to carry those letters past them. Girls scared me. This whole business scared me.

– Interview with Charles Weiner, March 5, 1966 (Niels Bohr Library and Archives with the Center for the History of Physics)

[On his first wife:] This probably happens to everybody, but anyway to me it seemed like independent and personal, that her feminine softness and different view of the world — and she was an artist, too — of what was valuable, what was beautiful, and so on, were things that I didn't ordinarily have direct interest in — like the lack of interest in humanities, in a way. But because of her interest in these things and the love that was developing between us, I paid a lot of attention to these matters, and softened up. I became a better guy as a result of the relationship and of listening to her ideas.

– Interview with Charles Weiner, March 5, 1966 (Niels Bohr Library and Archives with the Center for the History of Physics)

Philosophy and Religion

Philosophers say a great deal about what is absolutely necessary for science, and it is always, so far as one can see, rather naive, and probably wrong.

– *Feynman Lectures on Physics*, pp. 2–7

The science creates a power through its knowledge, a power to do things: You are able to do things after you know something scientifically. But the science does not give instructions with this power as to how to do some good as against how to do evil.

– Galileo Symposium, "What Is and What Should Be the Role of Scientific Culture in Modern Society," September 1964

The question enters many people's minds as to whether they will ever understand something which they like to call subjective consciousness or a personal conscience.

– UC Berkeley Lectures, "Time and Physics in Evolutionary History," spring 1968

How can we maintain that inspiration without a metaphysical theory, without necessarily believing certain ideas, such as that Christ rose from the dead, or something like that. Is that necessary in order to believe that the idea that you should help your neighbor, or that you should do unto your neighbor as he does to you; is it necessary to believe that Christ rose from the dead, in order to live like a Christian?

– Interview for Viewpoint

Nevertheless, a very great deal more truth can become known than can be proven.

– From *Nobel Lectures, Physics 1963–1970*, Elsevier Publishing Company, Amsterdam, 1972

I, a universe of atoms, an atom in the universe.

– "The Value of Science," December 1955

It doesn't seem to me that this fantastically marvelous universe, this tremendous range of time and space and different kinds of animals, and all the different planets, and all these atoms with all their motions and so on, all this complicated thing can merely be a stage so that God can watch human beings struggle for good and evil — which is the view that religion has. The stage is too big for the drama.

– Interview for Viewpoint

I would say the world-view of the physicists is not encroaching on any of the assumptions of biology or religion or whatever.

– UC Berkeley Lectures, "Time and Physics in Evolutionary History," spring 1968

You cannot define *anything* precisely. If we attempt to, we get into that paralysis of thought that comes to philosophers who sit opposite each other, one saying to the other, "you don't know what you are talking about!" The second one says, "What do you mean by *know*? What do you mean by *talking*? What do you mean by *you*?" and so on.

– *Feynman Lectures on Physics*, Lecture 8, October 20, 1961

I really didn't intend to insist that ethics and science are separate, but rather that the fundamental basis of ethics must be chosen in some non-scientific way. Then, when this is chosen, of course, science can help to decide whether we should or should not do certain things. Science can help us see what might happen if we do them, but the question as to whether we want something to happen depends on a choice of the ultimate ethical good.

– Letter to Professor Lawrence Cranberg, March 1965

But it turns out that falsehood and evil can be taught as easily as good.

– "The Uncertainty of Values," John Danz Lecture Series, 1963

Peace is a great force for good or for evil. How it will be for evil, I do not know. We will see, if we ever get to peace.

– "The Uncertainty of Values," John Danz Lecture Series, 1963

Scientists are explorers, philosophers are tourists.

– *No Ordinary Genius*, p. 260

The proponents of one idea have looked with horror at the actions of the believers of another. Horror because from a disagreeing point of view all the great potentialities of the race were being channeled into a false and confining alley. In fact, it is from the history of the enormous monstrosities that have been created by the false belief that philosophers have come to realize the fantastic potentialities and wondrous capacities of human beings.

– "The Uncertainty of Values," John Danz Lecture Series, 1963

Looking back at the worst times, it always seems that they were times in which there were people who believed with absolute faith and absolute dogmatism in something. And they were so serious in this matter that they insisted that the rest of the world agree with them.

– "The Uncertainty of Values," John Danz Lecture Series, 1963

I say that we do not know what is the meaning of life and what are the right moral values, that we have no way to choose them.

– "The Uncertainty of Values," John Danz Lecture Series, 1963

All I am trying to do is cast some doubt of confusion into the principle that survival permits ethics without question and that all people will agree that survival is the real determinate of good. If you can see that there may be some doubt about that, who would resolve the doubt for science?

– Letter to Professor Lawrence Cranberg, March 1965 (*Perfectly Reasonable Deviations from the Beaten Track*, p. 150)

Well, then how about the vocabulary. Have we got too many words? No, no. We need it to express ideas. Have we got too few words? No. By some accident, of course, through the history of time, we happened to have developed by accident the perfect combination of words.

– "The Unscientific Age," John Danz Lecture Series, 1963

This problem of moral values and ethical judgments is one into which science cannot enter.

– "The Unscientific Age," John Danz Lecture Series, 1963

To see generosity you must be generous enough not to see the meanness, and to see just meanness in a man you must be mean enough not to see the generosity.

– Letter to Reverend John Alex and Mrs. Marjorie Howard, December 1965 (*Perfectly Reasonable Deviations from the Beaten Track*, p. 184)

If you thought you were trying to find out more about it because you're going to get an answer to some deep philosophical question, you may be wrong. It may be that you can't get an answer to that particular question by finding out more about the character of nature. My interest in science is simply to find out about the world.

– BBC, "The Pleasure of Finding Things Out," 1981

I do not wish to accept membership in the American Philosophical Society. My reasons are entirely personal and reflect in no way on my opinion of the society. On the contrary, I feel it is a fine organization.

– Letter to George W. Corner and the American Philosophical Society, July 1968

What, then, is the meaning of it all? What can we say today to dispel the mystery of existence?

– "The Uncertainty of Values," John Danz Lecture Series, 1963

I agree that science cannot disprove God. I absolutely agree. I also agree that a belief in science and religion is consistent. I know many scientists who believe in God. It is not my purpose to disprove anything.

– "The Uncertainty of Values," John Danz Lecture Series, 1963

It seems to me that there is a kind of independence between the ethical and moral views and the theory of the machinery of the universe.

– "The Uncertainty of Values," John Danz Lecture Series, 1963

Because a little knowledge is dangerous, that the young man just learning a little science thinks he knows it all.

– "The Uncertainty of Values," John Danz Lecture Series, 1963

It seems that the metaphysical aspects of religion have nothing to do with the ethical values, that the moral values seem somehow to be outside of the scientific realm.

– "The Uncertainty of Values," John Danz Lecture Series, 1963

Men, philosophers of all ages, have tried to find the secret of existence, the meaning of it all. Because if they could find the real meaning of life, then all this human effort, all this wonderful potentiality of human beings, could then be moved in the correct direction and we would march forward with great success. So therefore we tried these different ideas. But the question of the meaning of the whole world, of life, and human beings, and so on, has been answered very many times by very many people. Unfortunately all the answers are different, and the people with one answer look with horror at the actions and behavior of the people with another answer.

– Galileo Symposium, "What Is and What Should Be the Role
   of Scientific Culture in Modern Society," September 1964

We do not know what the meaning of existence is. When we say at the result of studying all of the views that we have had before we find that we do not know the meaning of existence; but in saying that we do not know the meaning of existence, we have probably found the open channel.

– Galileo Symposium, "What Is and What Should Be the Role of
   Scientific Culture in Modern Society," September 1964

What science is, is not what the philosophers have said it is.

– National Science Teachers Association Fourteenth Convention
 lecture, "What Is Science?" April 1966

In religion, the moral lessons are taught, but they are not just taught once — you are inspired again and again, and I think it is necessary to inspire again and again, to remember the value of science for children, for grownups, and everybody else, in several ways.

– National Science Teachers Association Fourteenth Convention
 lecture, "What Is Science?" April 1966

A man cannot live beyond the grave.

– National Science Teachers Association Fourteenth Convention
 lecture, "What Is Science?" April 1966

Now, we know that, even with moral values granted, human beings are very weak; they must be reminded of the moral values in order that they may be able to follow their consciences. It is not simply a matter of having a right conscience; it is also a question of maintaining strength to do what you know is right. And it is necessary that religion give strength and comfort and the inspiration to follow these moral views. This is the inspirational aspect of religion. It gives inspiration not only for moral conduct — it gives inspiration for the arts and for all kinds of great thoughts and actions as well.

– "The Relation of Science and Religion," May 1956

It is a great adventure to contemplate the universe beyond man, to think of what it means without man — as it was for the great part of its long history, and as it is in the great majority of places. When this objective view is finally attained, and the mystery and majesty of matter are appreciated, to then turn the objective eye back on man viewed as matter, to see life as part of the universal mystery of greatest depth, is to sense an experience which is rarely described. It usually ends in laughter, delight in the futility of trying to understand. These scientific views end in awe and mystery, lost at the edge in uncertainty, but they appear to be so deep and so impressive that the theory that it is all arranged simply as a stage for God to watch man's struggle for good and evil seems to be inadequate.

 – "The Relation of Science and Religion," May 1956

Although I believe that from time to time scientific evidence is found which may be partially interpreted as giving some evidence of some particular aspect of the life of Christ, for example, or of other religious metaphysical ideas, it seems to me that there is no scientific evidence bearing on the golden rule. It seems to me that that is somehow different.

 – "The Relation of Science and Religion," May 1956

Western civilization, it seems to me, stands by two great heritages. One is the scientific spirit of adventure — the adventure into the unknown, an unknown which must be recognized as being unknown in order to be explored; the demand that the unanswerable mysteries of the universe remain unanswered; the attitude that all is uncertain; to summarize it — the humility of the intellect. The other great heritage is Christian ethics the basis of action on love, the brotherhood of all men, the value of the individual — the humility of the spirit. These two heritages are logically, thoroughly consistent. But logic is not all; one needs one's heart to follow an idea.

– "The Relation of Science and Religion," May 1956

One thing that Von Neumann gave me was an idea that he had which was interesting. That you don't have to be responsible for the world that you're in, and so I have developed a very powerful sense of social irresponsibility as a result of Von Neumann's advice. It's made me a very happy man since.

– UCSB talk, "Los Alamos from Below," February 1975

The uncertainty that is necessary in order to appreciate nature is not easily correlated with the feeling of certainty of faith which is usually associated with deep religious belief.

– "The Uncertainty of Values," John Danz Lecture Series, 1963

Most people are bad, you know, in one way or another — but they are not always bad, and they have other good points to compensate.

– Correspondence with Mimi Phillips, published in the *Wheeling News Register*, October 5, 1958

At thirteen I was not only converted to other religious views but I also stopped believing that the Jewish people are "the chosen people."

– Letter to Tina Levitan, February 1967 (*Perfectly Reasonable Deviations from the Beaten Track*, p. 236)

We have everybody here who's coming, I guess, but I see somebody that isn't here — how do you like that? Now that's an old philosophical question, isn't it, whether you can see someone that isn't here? I remember arguing at Princeton with the graduate students for hours and hours — the philosophy students as to what you were talking about when you said there was no chicken in the icebox. That's why I have nothing to do with philosophers.

– Esalen lecture, "Quantum Mechanical View of Reality (Part 1)," October 1984

In my opinion, it is not possible for religion to find a set of metaphysical ideas which will be guaranteed not to get into conflicts with an ever advancing and always changing science which is going into an unknown. We don't know how to answer the questions; it is impossible to find an answer which someday will not be found to be wrong. The difficulty arises because science and religion are both trying to answer questions in the same realm here.

– "The Relation of Science and Religion," May 1956

So I often wonder: What is the relation of integrity to working in the government?

– *What Do You Care What Other People Think?*, p. 218

One thing is that I have no more stomach for philosophical questions and political questions. One of the reasons is I got — I shy away from them much more positively than I ever did before. I won't even discuss them. I don't know why.

– Interview with Charles Weiner, June 28, 1966 (Niels Bohr Library and Archives with the Center for the History of Physics)

Why make yourself miserable saying things like, "Why do we have such bad luck? What has God done to us? What have we done to deserve this?" all of which, if you understand reality and take it into your heart, are irrelevant and unsolvable. They are just things that nobody can know. Your situation is just an accident of life.

– *What Do You Care What Other People Think?*, p. 51

In this age of specialization, men who thoroughly know one field are often incompetent to discuss another. The great problems of the relations between one and another aspect of human activity have for this reason been discussed less and less in public. When we look at the past great debates on these subjects, we feel jealous of those times, for we should have liked the excitement of such argument. The old problems, such as the relation of science and religion, are still with us, and I believe present as difficult dilemmas as ever, but they are not often publicly discussed because of the limitations of specialization.

– "The Relation of Science and Religion," *Engineering and Science*, May 1956

Nature of Science

You ask, are we getting anywhere? I'm reminded of a situation when I was asked the same thing. I was trying to pick a safe. Somebody asked me, "How are you doing? Are you getting anywhere? You can't tell until you open it. But you have tried a lot of numbers that you know don't work.

> – Panel discussion, particle physics conference, Irvine, California, 1971

[On the relation of physics to other sciences:] Anyway, you see it's all so interconnected it's not worth calling them different names except for convenience.

– Audio recording of Feynman Lectures on Physics, Lecture 3,
October 3, 1961, Q&A

For today all physicists know, from studying Einstein and Bohr, that sometimes an idea which looks completely paradoxical at first, if analyzed to completion in all detail and in experimental situations, may, in fact, not be paradoxical.

– From *Nobel Lectures, Physics 1963–1970*, Elsevier Publishing
Company, Amsterdam, 1972

I am familiar with a number of experimental physicists and they are sort of men of the earth. Therefore, I have always suspected that, one day, working far away from theorists, close to their big machines, they will get the idea of a new experiment: an experiment which will test the oracle. They would like to see what would happen, just for the fun of it, if they falsely report that there exists a certain bump, or an oscillation in a certain curve, and see how the theorists predict it. I know these men so well that the moment I thought of that possibility I have honestly always been concerned that some day they will do just that. That you can imagine how absurd the theoretical physicists would sound, making all these complicated calculations to demonstrate the existence of such a bump, while these fellows are laughing up their sleeves. For these reasons, I have found myself almost incapable of making calculations of the type that most other people make. I am afraid they will catch me out!

– Programme of American Physical Society Annual Meeting, 1950

There's so much distance between the fundamental rules and the final phenomena that it's unbelievable that the final variety of phenomena can come from such a steady operation of simple rules.

– Interview with Yorkshire Television program, "Take the World from Another Point of View," 1972

Scientific knowledge enables us to do all kinds of things and to make all kinds of things. Or course if we make good things, it is not only to the credit of science; it is also to the credit of the moral choice which led us to good work. Scientific knowledge is an enabling power to do either good or bad — but it does not carry instructions on how to use it.

– "The Value of Science," December 1955

We like to say that the world outside is real or what we can measure is real and if you look at it a bit longer you realize that the only thing that is real is what you felt you measured and how you feel that that the outside world could have easily been an illusion of the brain.

– UC Berkeley Lectures, "Time and Physics in Evolutionary History," spring 1968

A scientific argument is likely to involve a great deal of laughter and uncertainty on both sides, with both sides thinking up experiments and offering to bet on the outcome.

– "The Uncertainty of Science," John Danz Lecture Series, 1963

It's one of the advantages of modern society: that we don't have to cope with these difficult technical problems.

> – Audio recording of lecture on relativity, Douglas Advanced Research Laboratory, 1967

But there is nothing in biology yet found that indicates the inevitability of death.

> – Galileo Symposium, "What Is and What Should Be the Role of Scientific Culture in Modern Society," September 1964

Now you can see why it is I feel a bit uncomfortable when someone asks to give a talk: "Please tell us the latest things," because then, of course, I'm talking about the problems we have in understanding the insides of a proton! So, that gives us some idea as to why it is I feel so uncomfortable talking all the time about things we don't know much about, and nobody asks me about the stuff we do know everything about!

> – "QED: Photons — Corpuscles of Light," Sir Douglas Robb Lectures, University of Auckland, June 1979

The only way to have real success in science, the field I'm familiar with, is to describe the evidence very carefully without regard to the way you feel it should be. If you have a theory, you must try to explain what's good and what's bad about it equally. In science, you learn a kind of standard integrity and honesty.

> – *What Do You Care What Other People Think?*, pp. 117–118

You say, the trouble with the teachers of physics is that they don't explain quantum mechanics like this, that the particles behave like waves, or they behave like billiard balls, or they behave like — I don't know what. They don't behave like anything that you know of, you see. So it's impossible to describe the things except in analytic ways.

– Audio recording of Feynman Lectures on Physics, Lecture 2, September 29, 1961

As usual, nature's imagination far surpasses our own, as we have seen from the other theories which are subtle and deep. To get such a subtle and deep guess is not so easy. One must be really clever to guess, and it is not possible to do it blindly by machine.

– *The Character of Physical Law*, p. 162

Yeah, well, you shine a light in the corner and that's exciting when you run in there but if you get stuck to a wall then there's no use — you have to find another way out, right?

– Interview with Charles Weiner, February 4, 1973 (Niels Bohr Library and Archives with the Center for the History of Physics)

Physics is not mathematics, and mathematics is not physics. One helps the other. But in physics you have to have an understanding of the connection of words with the real world. It is necessary at the end to translate what you have figured out into English, into the world, into the blocks of copper and glass that you are going to do the experiments with. Only in that way can you find out whether the consequences are true.

– *The Character of Physical Law*, p. 55

Now this power to do things carries with it no instructions on how to use it, whether to use it for good or for evil. The product of this power is either good or evil, depending on how it is used.

– "The Uncertainty of Science," John Danz Lecture Series, 1963

In general, we look for a new law by the following process. First we guess it. Then we compute the consequences of the guess to see what would be implied if this law that we guessed is right. Then we compare the result of the computation to nature, with experiment or experience, compare it directly with observation, to see if it works. If it disagrees with experiment it is wrong. In that simple statement is the key to science. It doesn't make any difference how beautiful your guess is. It does not make any difference how smart you are, who made the guess, or what his name is — if it disagrees with experiment it is wrong.

– *The Character of Physical Law*, p. 156

The mathematicians are exploring in all directions, and it's quicker for a physicist to catch up on what he needs than to try to keep up with everything that might conceivably be useful.

– *Omni* interview, February 1979

We are trying to prove ourselves as wrong as quickly possible, because only in that way can we find progress.

– *The Character of Physical Law*, p. 158

Discovering the laws of physics is like trying to put together the pieces of a jigsaw puzzle. We have all these different pieces, and today they are proliferating rapidly. Many of them are lying about and cannot be fitted with other ones. How do we know that they belong together? How do we know that they are really all part of one picture? We are not sure, and it worries us to some extent, but we get encouragement from the common characteristics of several pieces. They all show blue sky, or they are all made out of the same kind of wood.

– *The Character of Physical Law*, p. 83

What is necessary "for the very existence of science," and what the characteristics of nature are, are not to be determined by pompous preconditions, they are determined by the material with which we work, by nature herself. We look, and we see what we find, and we cannot say ahead of time successfully what it is going to look like.

– *The Character of Physical Law*, p. 147

If science is to progress, what we need is the ability to experiment, honesty in reporting results — the results must be reported without somebody saying what they would like the results to have been — and finally — an important thing — the intelligence to interpret the results.

– *The Character of Physical Law*, p. 148

So my antagonist said, "Is it impossible that there are flying saucers? Can you prove that it's impossible?" "No," I said. "I can't prove that it's impossible. It's just very unlikely." At that he said, "You are very unscientific. If you can't prove it impossible then how can you say that it's unlikely?" But that is the way that is scientific. It is scientific only to say what is more likely and what is less likely, and not to be proving all the time the possible and impossible.

– *The Character of Physical Law*, pp. 165–166

When you're putting ideas together which are vague and hard to remember, it's like — I get this feeling very much — it's like building those houses of cards, and each of the cards is shaky, and if you forget one of them, the whole thing collapses and you don't know how you got there. You have to build them up again. If you are interrupted, and you forget how the cards go together (the cards being different parts of the idea, or different kinds of idea that have to go together to build up the main idea), it's quite a tower, and it's easy to slip. It needs a lot of concentration — solid time to think.

– BBC, "The Pleasure of Finding Things Out," 1981

It's a much deeper and warmer understanding, and it means you can be digging somewhere where you're temporarily convinced you'll find the answer, and somebody comes up and says, "Have you seen what they're coming up with over there?", and you look up and say, "Jeez! I'm in the wrong place!" It happens all the time.

– *Omni* interview, February 1979

The idea is to try to give all the information to help others to judge the value of your contribution: not just the information that leads to judgment in one particular direction or another.

– "Cargo Cult Science," Caltech commencement address, 1974

Scientific integrity. A principle of scientific thought that is utter honesty.

– Notes

The principle of science, the definition almost, is the following: the test of all knowledge is experiment. Experiment is the sole judge of "proof."

– Audio recording of Feynman Lectures on Physics, Lecture 1, September 26, 1961

In this complicated array of moving things, you can imagine it's something like trying to watching a great chess game being played by the gods. And you don't know what the rules of the game are, but all you're allowed to do is watch the playing. Now of course if you watch long enough, you may eventually catch on to a few of the rules. The rules of the game is what I mean by fundamental physics.

– Audio recording of Feynman Lectures on Physics, Lecture 2, September 29, 1961

If they tell us that a certain situation must always produce the same result, that's all very well. But if when we try it and it doesn't occur, then it doesn't occur. We just have to take what we see.

– Audio recording of Feynman Lectures on Physics, Lecture 2, September 29, 1961

Because science is good, let's say, it doesn't mean that which is not science is not good.

– Audio recording of Feynman Lectures on Physics, Lecture 3, October 3, 1961

Incidentally, psychoanalysis is not a science: it's a medical business, it's like witch doctory. It has a theory as to what causes disease — lots of different spirits, and this and that. Let's say the witch doctor has a theory that the disease is caused by a spirit which comes through the air; it's true, something does come in through the air — but it's not the same kind of spirit, and it is not gotten rid of say, by shaking a snake over it, but quinine does help malaria. So, if you're sick, I would advise that you go to the witch doctor because he is the man in the tribe who knows the most about the disease; on the other hand, it is not a science.

– Audio recording of Feynman Lectures on Physics, Lecture 3, October 3, 1961

It's easy enough to make up a theory by talking.

– Audio recording of Feynman Lectures on Physics, Lecture 7, October 17, 1961, Q&A

It wasn't that I knew that I was smart. It was just that — you see, scientific things are rationally right.

– Interview with Charles Weiner, 1966, March 4, 1966 (Niels Bohr Library and Archives with the Center for the History of Physics)

Yeah, I'm always interested in the relationship, the practical business — that the thing does not mean anything really unless there's some way to make something go.

> – Interview with Charles Weiner, March 4, 1966 (Niels Bohr
> Library and Archives with the Center for the History of Physics)

You must know, if you know how to play chess, that it's easy to learn all the rules, and yet it's very hard to select the best move, or to understand why Alekhine did that. And so, in the same manner, only much worse is it in nature, that we may be able to find all the rules — actually, we do not have all the rules, we know that we don't have all the rules — every once something like "castling" or something is going on, that we still don't understand.

> – Audio recording of Feynman Lectures on Physics, Lecture 2,
> September 29, 1961

Because there are many people who feel that the way to do it is to find out what problem the big guys are working on, and work right there. It isn't such a good scheme. You can be chasing the wrong thing, if you haven't thought about it yourself.

> – Interview with Charles Weiner, March 5, 1966 (Niels Bohr
> Library and Archives with the Center for the History of Physics)

If it's physics, it's interesting. See, everything fits together.

> – Interview with Charles Weiner, March 5, 1966 (Niels Bohr
> Library and Archives with the Center for the History of Physics)

But the beautiful experiment, which I still remember, in the lab, was this. There was a ring. You know, other experiments — dropping with the apparatus, with spark caps, with wheels, with all kinds of things. There was a hook on the wall; I mean a nail driven into the wall, and a ring of metal, a metallic ring, an amelus, whatever you call it, like a big washer, a big thing. It said: "Hang on the wall, measure the period, calculate the period from the shape, and see if they agree." I loved that. I thought that was the best doggoned thing. I didn't care as much — I'm just trying to remember now — I liked the other experiments, but they involved the sparks and all the other hocus pocus, which was too easy. With all that equipment, you could measure the acceleration of gravity. But to think that physics is so good, not that you can figure out something carefully prepared, but something as natural as a lousy old ring hanging off a hook — it impressed me, that I had now the power to tell what something as dumb as that was going to do.

> – Interview with Charles Weiner, March 5, 1966 (Niels Bohr Library and Archives with the Center for the History of Physics)

The situation is, as it always is when we're near the answer, it looks much simpler than it has any right to be. And we have to understand that simplicity. And why we think it must be more complicated. Our minds are too complicated, somehow.

> – Yorkshire Television interview, "Take the World from Another Point of View," 1972

I have a principle in strong interaction theories; if the theory is complicated, it's wrong.

– Panel discussion, particle physics conference, Irvine, California, 1971

I'm not at all pessimistic that the answer will ultimately fall out. You can't fail. Nature cannot resist the perpetual penetration by experiment, and those apes which sit crudely and clumsily watching what's happening, sooner or later are going to see, after their noses are rubbed in the answer, what it is.

– Panel discussion, particle physics conference, Irvine, California, 1971

You can recognize truth by its beauty and simplicity.

– *The Character of Physical Law*, p. 171

Some theorists do not understand the relation of theory and experiment — and where lies the true source and test of all their knowledge?

– Letter to Dr. Blas Cabrera, September 1982 (*Perfectly Reasonable Deviations from the Beaten Track*, p. 349)

In science (unlike business or any other pursuits) we are all working together cooperating to try to understand Nature and we have learned to be very careful to recognize and commend anybody who gets a really useful new idea.

– Letter to Dr. Rafael Dy-Liacco, June 1978 (*Perfectly Reasonable Deviations from the Beaten Track*, p. 321)

New ideas are always fascinating, because physicists wish to find out how Nature works. Any experiment which deviates from expectations according to known law commands immediate attention, because we may find something new.

– Letter to Mr. L. Dembart (*Los Angeles Times*), January 1986 (*Perfectly Reasonable Deviations from the Beaten Track*, p. 397)

Most of our problems are man-made, of this type.

– Audio recording of lecture on relativity, Douglas Advanced Research Laboratory, 1967

In fact, an engineer is probably best defined as a man who knows how to convert energy from one unit to another.

– Audio recording of lecture on relativity, Douglas Advanced Research Laboratory, 1967

The complications of relativistic arguments is that you have two points of view. The common thing that a novice does is shift back and forth between the two during a calculation.

– Audio recording of lecture on relativity, Douglas Advanced Research Laboratory, 1967

It is amazing how little predictive ability theoretical physics has, even when it relies solely upon well-established fundamental principles.

– Letter to Dr. Hans Bethe, June 1951

There is today, in my opinion, no science capable of adequately selecting or judging people. So I doubt that any intelligent method is known.

– Letter to Mr. Douglas M. Fowle, September 1962 (*Perfectly Reasonable Deviations from the Beaten Track*, p. 135)

Physicists looking at psychology, through force of habit, try to suggest some simple element be found to study, rather than the whole human brain at once.

– On human and animal eyesight, in a letter to Edwin H. Land (Polaroid Corporation), May 1966 (*Perfectly Reasonable Deviations from the Beaten Track*, p. 223)

More and more people find fundamental physics a relatively uninteresting subject. So it is left in an incomplete state, with a few working very slowly at the edge on the question of what is the third order tensor field that has a coupling constant 10^{30} times smaller than gravity?

– MIT centennial, "Talk of Our Times," December 1961

The purpose of scientific thought is to predict what will happen in given experimental circumstances.

– Letter to F. Harrison Stamper, April 1962

It is not always a good idea to be too precise.

– "The Uncertainty of Science," John Danz Lecture Series, 1963

The most obvious characteristic of science is its application, the fact that as a consequence of science one has a power to do things.

– "The Uncertainty of Science," John Danz Lecture Series, 1963

But if a thing is not scientific, if it cannot be subjected to the test of observation, this does not mean that it is dead, or wrong, or stupid. We are not trying to argue that science is somehow good and other things are somehow not good.

– "The Uncertainty of Science," John Danz Lecture Series, 1963

Incidentally, the fact that there are rules at all to be checked is a kind of miracle; that it is possible to find a rule, like the inverse square law of gravitation, is some sort of miracle.

– "The Uncertainty of Science," John Danz Lecture Series, 1963

Science is not a specialist business; it is completely universal.

– "The Uncertainty of Science," John Danz Lecture Series, 1963

The speed at which science has been developing for the last 200 years has been ever increasing, and we reach a culmination of speed now.

– "The Unscientific Age," John Danz Lecture Series, 1963

That something is unscientific is not bad, that is nothing the matter with it.

– "The Unscientific Age," John Danz Lecture Series, 1963

Unlike other scientific subjects, there has not been a general detailed development of our ideas of time. We do not know a great deal more than is known from common sense. Everyone knows how it flows inexorably onward and the scientist in all his studies has found relatively little to add to our knowledge of that mystery.

– From notes for "About Time" program, 1957

Scientific problems like to separate into two kinds. One, to understand the origin of the arrangement of things as we find them, and two, to understand how they will behave if started in a given circumstance.

– From notes for "About Time" program, 1957

It is interesting that this thoroughness, which is a virtue, is often misunderstood. When someone says a thing has been done scientifically, often all he means is that it has been done thoroughly.

– "The Uncertainty of Science," John Danz Lecture Series, 1963

Suppose Galileo were here and we were to show him the world today and try to make him happy, or see what he finds out. And we would tell him about the questions of evidence, those methods of judging things which he developed. And we would point out that we are still in exactly the same tradition, we follow it exactly — even to detail of making numerical measurements and using those as one of the better tools, in the physics at least. And that the sciences have developed in a very good way

directly and continuously from his original, in the same spirit
as he developed. And as a result there are no more witches and
ghosts.

– Galileo Symposium, "What Is and What Should Be the Role of
Scientific Culture in Modern Society," September 1964

Now, it might be true that astrology is right. It might be true that
if you go to the dentist on the day that marks as at right angles to
Venus, that it is better than if you go on a different day. It might
be true that you can be cured by the miracle of Lourdes. But if
it is true it ought to be investigated. Why? To improve it. If it is
true then maybe we can find out if the stars do influence life;
that we could make the system more powerful by investigating
statistically, scientifically judging the evidence and objectively
more carefully.

– Galileo Symposium, "What Is and What Should Be the Role of
Scientific Culture in Modern Society," September 1964

Science is a long history of learning how not to fool ourselves.

– Quoted in K. C. Cole, *The Universe and the Teacup: The Mathematics of Truth and Beauty*, 1998

I believe that we must attack these things in which we do not
believe. Not attack by the method of cutting off the heads of the
people but attack in the sense of discuss.

– Galileo Symposium, "What Is and What Should Be the Role of
Scientific Culture in Modern Society," September 1964

I am thoroughly convinced that the world-view of physics is adequate to describe the phenomena of experiences of other sciences.

– UC Berkeley Lectures, "Time and Physics in Evolutionary History," spring 1968

In physics we believe very nicely that we have a theory by which if you tell us the initial conditions, we will predict what will happen in the future.

– UC Berkeley Lectures, "Time and Physics in Evolutionary History," spring 1968

Other theories, subjects, have a thing called the past as a fundamental part of that theoretical structure. But physics does not usually require a study of the past in order to analyze what is going to happen next.

– UC Berkeley Lectures, "Time and Physics in Evolutionary History," spring 1968

There is a way of thinking about the past that makes it the present, which makes a theory of the past really a predictive theory as to what will happen in the future.

– UC Berkeley Lectures, "Time and Physics in Evolutionary History," spring 1968

There is this relation of physics and biology in which we all talk so much and we try to describe the biological world always describing things outside. We always try to describe you, never me.

– UC Berkeley Lectures, "Time and Physics in Evolutionary History," spring 1968

If there is such a consciousness, then we have a large number of questions about how did it evolve, how extensive is it, how spread is it throughout the natural world among other animals or even the natural world and so on. It produces more questions than answers. Nevertheless there are many of you I suspect who have some kind of a feeling that everything is real from the inside as it is in a light.

– UC Berkeley Lectures, "Time and Physics in Evolutionary History," spring 1968

We're not trying to find out for certainty what the world looks like but just what we've seen so far.

– Esalen lecture, "Quantum Mechanical View of Reality (Part 1)," October 1984

We suspend an assumption that we are able to say that "this is an either this or that condition" when you haven't made any observation yet. And then we can continue to think straightforwardly, and that's the most efficient way we know.

– Esalen lecture, "Quantum Mechanical View of Reality (Part 1)," October 1984

That is what science is; the result of the discovery that is worthwhile rechecking by new direct experience, and not necessarily trusting the race experience from the past.

– National Science Teachers Association Fourteenth Convention
   lecture, "What Is Science?" April 1966

There is also a value of science in the value which it teaches the value of rational thought, as well as the importance of freedom of thought; the positive results that come from doubting that the lessons are true.

– National Science Teachers Association Fourteenth Convention
   lecture, "What Is Science?" April 1966

Usually when we publish things, things that we publish in the technical journals are very carefully polished and all of the routes and side alleys and back thoughts and so on that you had are taken out. You do not describe your personal adventure and succession of ideas. Any references that are given are given to the first man who thought of the thing, not the guy who told you about it, and so on.

– CERN talk, December 1965

I think that maybe that's why young people make success. They don't know enough. Because when you know enough it's obvious that every idea that you have is no good.

– CERN talk, December 1965

All they're doing is they're taking the physical world and they're slicing it up to analyze it. And this one you slice a different way, and no matter which way you slice it, it's the same bologna, so you should look at the bologna, not the slice process.

– CERN talk, December 1965

Say, look, what a wild and wonderful tour I took to come back to nothing. I did not ever solve the problem of infinite self-energy; it's still just as infinite as it was before I began.

– CERN talk, December 1965

Each of the theories we have in the world seem to be possible to write in many different ways and to look at from very many different physical views.

– CERN talk, December 1965

It seems to me quite marvelous that the laws of physics are of such a kind that they can be written in another way that isn't obviously the same and yet is the same. I will, just for the fun of it, will propose — although I have to think about it more to see whether it's sensible — that that's because it's simple, and that a simple thing is something that an be described in two different ways and you don't know that you're describing the same thing. Whereas if you take a complicated thing, in order to fully describe it, you've said so much about it, that there's no other way that you could see that it looks different.

– CERN talk, December 1965

When you look at things from different physical points of view, if you'll not sell your soul to one of them, which is what I was doing, but take them all, then you're that much greater.

– CERN talk, December 1965

The odds that your theory will be in fact right and that the general thing that everybody's working on will be right, is low. But the odds that you, Little Boy Schmidt, will be the guy who figures a thing out, is not smaller.

– CERN talk, December 1965

It's very important that we do not all follow the same fashion. Because although it's ninety percent sure that the answer lies over there, where [Murray] Gell-Mann is working, what happens if it doesn't, and if everybody's doing this?

– CERN talk, December 1965

If you've made up your mind to test a theory, or you want to explain some idea, you should always decide to publish it whichever way it comes out. If we only publish results of a certain kind, we can make the argument look good. We must publish BOTH kinds of results.

– "Cargo Cult Science," Caltech commencement address, 1974

In spite of the fact that science could produce enormous horror in the world, it is of value because it can produce something.

– "The Value of Science," December 1955

Science is a way to teach how something gets to be known, what is not known, to what extent things are known (for nothing is known absolutely), how to handle doubt and uncertainty, what the rules of evidence are, how to think about things so that judgments can be made, how to distinguish truth from fraud, and from show.

These are certainly important secondary yields of teaching science, and physics in particular.

– "The Problem of Teaching Physics in Latin America," 1963

It might not be that way, in which case you're looking for whatever the hell it is that you find!

– *Omni* interview, February 1979

The whole idea you started with is gone! That's the kind of exciting thing that happens from time to time.

– *Omni* interview, February 1979

So far, physics has tried to find laws and constants without asking where they came from, but we may be approaching the point where we'll be forced to consider history.

– *Omni* interview, February 1979

The lesson you learn, as you grow older in physics is that what we can do is a very small fraction of what there is.

– *Omni* interview, February 1979

We have an illusion we can do any experiment we want. We all, however, come from the same universe evolving, and we don't really have any "real" freedom. For we obey certain laws and have come from a certain past.

– MIT conference, May 1981

There's also the possibility that you don't understand because you get a bit confused. You're sure that you must have misinterpreted what I said or something like that, and you get turned off. Let me assure you that most of the time you did interpret correctly what I said because it's going to be so shocking the way Nature actually works that you're not going to believe that either.

– "QED: Photons — Corpuscles of Light," Sir Douglas Robb Lectures, University of Auckland, June 1979

Science — pure science, that is — cannot flourish, or at best can only flourish by accident.

– Notes for talk on "Science in America"

It is natural to explain new ideas in terms of what is already in your own head, but all these concepts are piled on top of each other: this idea is taught in terms of that idea, and that idea is taught in terms of another idea, which comes from counting, which can be different for different people!

– *What Do You Care What Other People Think?*, p. 59

When the argument for the existence of a term is solely to explain a single experiment, it is a question of physical taste.

– Programme of American Physical Society Annual Meeting, 1950

At each meeting it always seems to me that very little progress is made. Nevertheless, if you look over any reasonable length of time, a few years say, you find a fantastic progress and it is hard to understand how that can happen at the same time that nothing is happening in any one moment. I think that it is something like the way clouds change in the sky — they gradually fade out here and build up there and if you look later it is different.

– Programme of American Physical Society Annual Meeting, 1950

It is clear to everyone today that physics is almost entirely in the hands of the experimenters. I think, nevertheless, that we should appreciate that theory is supposed to have a predictive value.

– Programme of American Physical Society Annual Meeting, 1950

If you try to test modern theory on its predictive value, you find that it is very weak.

– Programme of American Physical Society Annual Meeting, 1950

When a new particle or a new fact is discovered, notice that all the theorists do one of two things: they either form a group, or disperse.

– Programme of American Physical Society Annual Meeting, 1950

One should look very hard for an "expected" failure. I have probably been converted from my prejudice that it must fail, just in time to be caught off base by an experiment next month showing that it indeed does.

– "The Present Situation in Quantum Electrodynamics," Solvay conference, 1961

In the South Seas there is a cargo cult of people. During World War II, they saw airplanes land with lots of good material, and they want the same thing to happen now. So they've arranged to make things that look like runways, put fires along the sides, made a wooden hut for a man to sit in, with two wooden pieces on his head like headphones and bars of bamboo sticking out like antennas — he's the controller. They wait for the airplanes to land. They're doing everything right. The form is perfect. But it doesn't work. So I call these things "cargo cult science" because they follow all the apparent precepts and forms of scientific investigation, but they're missing something essential.

– *U.S. News and World Report* interview, February 1985

Scientific progress to discovering new things often results by noticing small deviations from the predictions of old theories, and this cannot be done unless the predictions are very precise and detailed.

– "The Qualitative Behavior of Yang–Mills Theory in 2+1 Dimensions," January 1981

If you've heard about the atom being a little solar system, the nucleus at the center like the Sun and the planets going around like electrons, then you're back in nineteen hundred and something.

– "QED: Electrons and Their Interactions," Sir Douglas Robb Lectures, University of Auckland, June 1979

It's interesting how some people find science so interesting, and other people find it kind of dull and difficult. Especially kids, some just eat it up.

– BBC, "Fun to Imagine" television series, 1983

In the case of science, I think that one of the things that make it difficult is that it takes a lot of imagination. It's very hard to imagine all the crazy things that things really are like.

– BBC, "Fun to Imagine" television series, 1983

These are kind of disciplines in the field of science where you have to learn to know when you know and when you don't know and what it is you know and what it is you don't know. You gotta be careful not to confuse yourself!

– Yorkshire Television program, "Take the World from Another Point of View," 1972

The world is strange; the whole universe is very strange, but when you look at the details, you find out that the rules are very simple.

– Yorkshire Television program, "Take the World from Another Point of View," 1972

I think that that the numbers are a problem in astronomy — the sizes and numbers.

– BBC, "Fun to Imagine" television series, 1983

Newton, I would say, is a genius about something — a teacher about something. He's the guy who taught us how to think about science in a modern way so that we could make some progress. He's the one who distinguished very carefully between the facts that he would develop and experimentally determine, "This really happened!"

– "QED: Fits of Reflection and Transmission," Sir Douglas Robb Lectures, University of Auckland, 1979

You first have to tell me why a gram is as big as it is. And it's because somebody chose a gram during the French Revolution or something. They decided that such-and-such is a gram, and then the electron is so many grams!

– "QED: New Queries," Sir Douglas Robb Lectures, University of Auckland, 1979

That is one advantage we physicists have; we know what we found out a century ago.

– Letter to Dr. Robert S. Alexander, November 1965

I wanted to give you some appreciation of the wonderful world and the physicist's way of looking at it, which, I believe, is a major part of the true culture of modern times. (There are probably professors of other subjects who would object, but I believe they are completely wrong.) Perhaps you will not only have some appreciation of this culture; it is even possible that you may want to join in the greatest adventure that the human mind has ever begun.

– *Feynman Lectures on Physics*, Appendix

The present concentration of research facilities and scientific universities has obvious dangers.

– Letter to Professor M. L. Oliphant, regarding expanding research overseas (*Perfectly Reasonable Deviations from the Beaten Track*, p. 82)

Many physicists are working very hard trying to put together a grand picture that unifies everything into one super-duper model. It's a delightful game, but at the present time none of the speculators agree with any of the other speculators as to what the grand picture is.

– *QED: The Strange Theory of Light and Matter*, p. 150

If I am gambling in Las Vegas, and am about to put some money into number twenty-two at roulette, and the girl next to me spills her drink because she sees someone she knows, so that I stop before betting, and twenty-two comes up, I can see that the whole course of the universe for me has hung on the fact that some little photon hit the nerve ends of her retina.

– *Feynman Lectures on Gravitation*

[On what would mark the end of physics:] Looking for trouble — and you don't find any.

– MIT panel

[On his father:] He did teach himself a great deal: He read a lot and studied a lot because, as I know now, he understood a great deal about science.

– Interview with Charles Weiner, March 4, 1966 (Niels Bohr Library and Archives with the Center for the History of Physics)

Science not purposeful — research in engineering is. Our greatest advances come from researchers not aimed at use but just for fun, curiosity, and desire for understanding.

– Notes from Los Alamos

No need to stop atomic research to control bomb. Only real requirement is to control use to which research is put.

– Notes from Los Alamos

Problems of chemistry and biology are greatly helped if our ability to see what we are doing and do things on an atomic level is ultimately developed — a development which I think cannot be avoided.

– Notes

Not one exact statement in science about anything.

– Notes

What happens when physical ideas get into the outside world, and are applied to society and psychology and so on, is that they become distorted to such a point that they become trivial, obvious, dull, things of no precision or accuracy.

– Audio recording of lecture on relativity, Douglas Advanced Research Laboratory, 1967

You gotta stop and think about it, to really get the pleasure about the complexity — the inconceivable nature of nature.

– BBC, "Fun to Imagine" television series, 1983

I think that the discovery of electricity and magnetism and the electromagnetic effects which are finally worked out are probably the most fundamental transformation of the most remarkable thing in history, the biggest change in history.

– BBC, "Fun to Imagine" television series, 1983

In order to talk about the impact of ideas in one field on ideas in another field one is always apt to be an idiot of one kind or another. In these days of specialization, there are few people who have such a deep knowledge of two departments of our understanding that they don't make fools of themselves in one or the other.

– "The Uncertainty of Science," John Danz Lecture Series, 1963

How can we say that only the best must be allowed in to join those who are already in, without loudly proclaiming to our inner selves that we who are in must be very good indeed. Of course I believe that I am very good indeed, but that is a private matter and I cannot publicly admit I do so, to such an extent that I have the nerve to decide that this man, or that, is not worthy of joining my elite club.

– Letter to Dr. Detlev W. Bronk and the National Academy of Sciences, August 1961 (*Perfectly Reasonable Deviations from the Beaten Track*, p. 108)

I introduced you to the gauge theories, if you recall. But I must confess I'm not at all happy about them. They're like maps showing the different particles. An ordinary map that shows the mountain peaks doesn't tell you why the mountains are there. You still have to find out how the Earth works.

– BBC interview, "Beyond Present Theories"

Science means, sometimes, a special method of finding things out. Sometimes it means the body of knowledge arising from the things found out. It may also mean the new things that you can do when you have found something out, or the actual doing of new things.

– "The Uncertainty of Science," John Danz Lecture Series, 1963

In learning science you learn to handle trial and error, to develop a spirit of invention and of free inquiry, which is of tremendous value far beyond science. One learns to ask oneself: "Is there a better way to do it?" (And the answer to this is not the conditioned reflex: "Let's see how they do it in the United States," because there must certainly be a better way than that!)

– "The Problem of Teaching Physics in Latin America," 1963

[On revelation:] Afterwards, you think "Why the devil was I so stupid and didn't see this?" Not only true of you but true of the history of science. You can always look at human history and wonder why they hadn't thought of it twenty years earlier or ten years earlier, depending on the pace.

– Yorkshire Television program, "Take the World from Another Point of View," 1972

I imagine experimental physicists must often look with envy at men like Kamerlingh Onnes, who discovered a field like low temperature, which seems to be bottomless and in which one can go down and down. Such a man is then a leader and has some temporary monopoly in a scientific adventure.

– "There's Plenty of Room at the Bottom," December 1959

Understanding space really means understanding how things might look from another point of view.

– Audio recording of lecture on relativity, Douglas Advanced Research Laboratory, 1967

In no field is all the research done. Research leads to new discoveries and new questions to answer by more research.

– Letter to student Mark Minguillon, August 1976 (*Perfectly Reasonable Deviations from the Beaten Track*, p. 306)

I believe most assuredly that the next science to find itself in moral difficulties with its applications is biology, and if the problems of physics relative to science seem difficult, the problems of the development of biological knowledge will be fantastic.

– Galileo Symposium, "What Is and What Should Be the Role of Scientific Culture in Modern Society," September 1964

The world goes around because of differences of opinion and interests resulting in a division of labor — even volunteer labor. I hope everyone does not think as I do — if I found out they did, I would change my view. For from this variety of approaches real progress must come. We must try everything.

– Letter to John M. Fowler, March 1966

I wouldn't say that my physics wasn't up to the prize, but I'm not up to it on a human side, being a Prize Winner and an Important Scientist. I'm not, that's all. I was a kid fooling around. I was in my pajamas working on the floor with paper and pencil and I cooked something up, OK?

– "The Remarkable Dr. Feynman," *Los Angeles Times Magazine*, April 20, 1986

So it's kind of fun to imagine that this intimate mixture of highly attractive opposites which are so strong that they cancel out the effects and it's only sometimes, when you have an excess of one kind or another that you get this mysterious electrical force. And how can I explain these electrical forces in any other way? Why should I try to explain it in terms something like jelly or other things which are made?

– BBC, "Fun to Imagine" television series, 1983

But we have to keep going, we find out more if we just keep going, so we keep going.

– Audio recording of Feynman Lectures on Physics, Lecture 7, October 17, 1961

Incidentally, may I say from the beginning that if a thing is not a science, it is not necessarily bad. For example, love is not a science. So, if something is said not to be a science, it does not mean that there is something wrong with it; it just means that it is not a science.

– Audio recording of Feynman Lectures on Physics, Lecture 3, October 3, 1961

I went down to the town to buy supplies and I was carrying a wastepaper basket and some other things, and [Leonard]

Eisenbud, who was a theoretical physicist — we met — he passed me on the street. "Ah," he said, "you look like you're going to be a good theoretical physicist. You've brought the right tools — there's an eraser and a wastebasket."

> – Interview with Charles Weiner, March 5, 1966 (Niels Bohr Library and Archives with the Center for the History of Physics)

[On Einstein:] He wore this sweater, without a shirt under it, no socks — just like everybody says — and was such a soft, nice man in the discussions, at all points. He was such an interesting man to talk to.

> – Interview with Charles Weiner, March 5, 1966 (Niels Bohr Library and Archives with the Center for the History of Physics)

You know that nature can look very, very strange, in the fundamentals, and yet produce in the end the natural phenomena in a way, very different looking than you would think at first. It's all right. You've got to think it out, you can't just jump to the conclusion that it's wrong.

> – Interview with Charles Weiner, March 5, 1966 (Niels Bohr Library and Archives with the Center for the History of Physics)

[On the first nuclear test:] I've always been impressed by acoustics. Acoustics have meaning for me. Not so much as the visual. When I heard the solidity of that crack, at 20 miles away, then I knew that that thing was something, and I got excited.

> – Interview with Charles Weiner, March 5, 1966 (Niels Bohr Library and Archives with the Center for the History of Physics)

I've found that my trouble is, I haven't been writing things up. I discover things, and then other people discover them later, and then it's not worth writing them up. For instance, I worked out the quantum theory of gravitation to an order infinitely higher — I mean, to a degree, to a detail, infinitely higher — than anybody else that I know. But it isn't complete. There are some slight weaknesses. So I haven't written it up. But it's crazy — that's five years old now. It should be written up.

– Interview with Charles Weiner, June 27, 1966 (Niels Bohr Library and Archives with the Center for the History of Physics)

What happens to it — this is interesting, this is the way science works, I found out — what happens to a person, when they believe something, is that they see a few that look like it, which are due to something else. Then they believe in the existence of the thing. Then all the rest of the things which are not good become corroborative evidence, not one bit of which is very strong, but seems to be in large amount. But the moment that you propose that it isn't true, all the corroborative evidence just disappears. I mean, it's just a very small business — it's just selection of a special plate, that looks like it's in the right direction. It is not strong. And so one can build up an argument and believe something, and think it has a large amount of weight, when actually if you go and look carefully at it the weight is very weak, and each item is weak and it doesn't add up to much.

– Interview with Charles Weiner, June 28, 1966 (Niels Bohr Library and Archives with the Center for the History of Physics)

I've always taken an attitude that I have only to explain the regularities of nature — I don't have to explain the methods of my friends. I don't have to learn the systems and methods of my friends — only the regularities of nature — and that's a very economizing way of proceeding.

– Interview with Charles Weiner, June 28, 1966 (Niels Bohr Library and Archives with the Center for the History of Physics)

Science is only useful if it tells you about some experiment that has not been done; it is no good if it only tells you what just went on.

– *The Character of Physical Law*, p. 164

Therefore psychologically we must keep all the theories in our heads, and every theoretical physicist who is any good knows six or seven different theoretical representations for exactly the same physics.

– *The Character of Physical Law*, p. 168

We are trained by Einstein and Bohr and so on, to appreciate that an idea which is prima facie absurd, can after time aligns enough, agree with experience. That is, things that look absolutely nutty, like lack of simultaneous time or uncertainty and so on, is perfectly possible.

– CERN talk, December 1965

We have a habit in writing articles published in scientific journals to make the work as finished as possible, to cover all the tracks, to not worry about the blind alleys or to describe how you had the wrong idea first, and so on.

> – From *Nobel Lectures, Physics 1963–1970*, Elsevier Publishing
> Company, Amsterdam, 1972

Research is a method of attaining information. You don't make a bomb by research — you find out how to do it by research.

> – Notes from Los Alamos

If you expect science to give you all the answers to the wonderful questions about what we are, where we are going, and what the meaning of the universe is, then I think you could easily become disillusioned and look for a mystic answer to these problems. How a scientist can want a mystic answer, I don't know.

> – BBC, "The Pleasure of Finding Things Out," 1981

And so it is with science. In a way it is a key to the gates of heaven, and the same key opens the gates of hell, and we do not have any instructions as to which is which gate. Shall we throw away the key and never have a way to enter the gates of heaven? Or shall we struggle with the problem of which is the best way to use the key?

> – "The Uncertainty of Science," John Danz Lecture Series, 1963

Control of scientific innovations means end of science. Science is source of all our engineering development — care not to kill it by control.

> – Notes from Los Alamos

Science alone of all subjects contains within itself the lesson of the danger of belief in the infallibility of the greatest teachers of the preceding generation.

— National Science Teachers Association Fourteenth Convention lecture, "What Is Science?" April 1966

Well, these scientific views end in awe and mystery, lost at the edge in uncertainty, but they appear to be so deep and so impressive that the theory that it is all arranged as a stage for God to watch man's struggle for good and evil seems inadequate.

— "The Uncertainty of Values," John Danz Lecture Series, 1963

Curiosity and Discovery

The more I ask why, it gets interesting. That's my idea, that the deeper a thing is the more interesting it gets.

– BBC, "Fun to Imagine" television series, 1983

I love challenges. I always have. In fact, later hobbies at the beginning were not scientific but were always challenges. Picking locks, cracking codes, analyzing hieroglyphics that nobody knows how to translate — you see, they're all the same now.

– Interview with Charles Weiner, March 4, 1966 (Niels Bohr Library and Archives with the Center for the History of Physics)

The other day, I was at the dentist, and he's getting ready with his electric drill to make holes, and I thought, I better think of something fast, or else it's gonna hurt! And then I thought about this little motor going around, and what was it that made it turn, and what was going on.

– BBC, "Fun to Imagine" television series, 1983

I think that to keep trying new solutions is the way to do everything.

– "The Uncertainty of Science," John Danz Lecture Series, 1963

I wonder why. I wonder why. I wonder why I wonder. I wonder *why* I wonder why I wonder why I wonder!

– *Surely You're Joking, Mr. Feynman!*, p. 48

The dream is to find the open channel.

– "The Uncertainty of Values," John Danz Lecture Series, 1963

What one fool can do, so can another, and the fact that some other fool beat you to it shouldn't disturb you: you should get a kick out of having discovered something.

– *Feynman Lectures on Computation*, pp. 15–16

Suppose people are exploring a new continent, OK? They see water coming along the ground, they've seen that before, and they call it "rivers." So they say they're exploring to find the headwaters, they go upriver, and sure enough, there they are, it's all going very well. But lo and behold, when they get up far enough they find the whole system's different: There's a great big lake, or springs, or the rivers run in a circle. You might say, "Aha! They've failed!" but not at all! The real reason they were doing it was to explore the land.

– *Omni* interview, February 1979

As long as I didn't know they discovered it, for that moment I knew something and I had found a law, and I could make predictions about nature, which is the aim that I had. And the fact that somebody else was already making the predictions, unbeknownst to me, in no way takes the pleasure away, in any way. So that was really a great moment. I would like more moments like that, but I don't have to ask the gods for everything.

– Interview with Charles Weiner, June 28, 1966 (Niels Bohr Library and Archives with the Center for the History of Physics)

Fiddling is the answer. Experimenting is fiddling around.

– Interview with Charles Weiner, March 5, 1966 (Niels Bohr
  Library and Archives with the Center for the History of
  Physics)

This stuff of fantasizing and looking at the world, imagining
things, which really isn't fantasizing because you're only trying
to imagine things the way it really is, comes in handy sometimes.

– BBC, "Fun to Imagine" television series, 1983

You come in to me now, in an interview, and you're asking me
about the latest discoveries that are made. Nobody every asks
about a simple, ordinary phenomenon in the street, you know, like,
"What about those colors?" Or something like that. We could have
a nice, interview, explain all about the colors. Butterfly wings and
whole big deal. Don't care about that. You want the big, final
result. Instead it's going to be complicated, because I am at the
end of 400 years of very effective method of finding things out
about the world.

– Yorkshire Television interview, "Take the World from Another
  Point of View," 1972

The truth always is funny in the sense that it explains a lot more
than you expected when you started to figure it out.

– Audio recording of Feynman Lectures on Physics, Lecture 7,
  October 17, 1961, Q&A

I can't understand anything in general unless I'm carrying along in my mind a specific example and watching it go.

– *Surely You're Joking, Mr. Feynman!*, p. 244

Man has been stopped before by stopping his ideas.

– "The Uncertainty of Values," John Danz Lecture Series, 1963

Furthermore, in the search for new laws, you always have the psychological excitement of feeling that possibly nobody has yet thought of the crazy possibility you are looking at right now.

– From *Nobel Lectures, Physics 1963–1970*, Elsevier Publishing Company, Amsterdam, 1972

The exceptions to any rule are most interesting in themselves, for they show us that the old rule is wrong. And it is most exciting then, to find out what the right rule is, if any, is.

– "The Uncertainty of Values," John Danz Lecture Series, 1963

I have to understand the world, you see.

– *Surely You're Joking, Mr. Feynman!*, p. 231

[On theoretical physics:] It's like exploring a new country and you cannot explore a country successfully if you care what you find because you might say to yourself "I don't think it's gonna come out the way I thought it was and so I'm gonna quit exploring." No, it's the fun of finding out what it's like rather than starting out with a pre-set idea and if you find that it isn't that way you get disappointed; that's a ridiculous thing, so I don't fill up my mind with ideas of which way it's going to go, I just try and find out as much as I can.

– BBC interview, "Scientifically Speaking," April 1976

We live in a heroic, a unique and wonderful age of excitement. It's going to be looked at with great jealousy in the ages to come. How would it have been to live in the time when they were discovering the fundamental laws? You can't discover America twice, and we can be jealous of Columbus. You say, yes, but if not America, then there are other planets to explore. That is true. And if not fundamental physics, then there are other questions to investigate.

– *Perfectly Reasonable Deviations from the Beaten Track*, p. 440

And it is a miracle that it's possible, by doing experiments over here, to predict what's going to happen over there. It is not as much a miracle to predict something if you know the laws about it. In other words, it's enough of a miracle that there are laws at all, but what's really a miracle is to be able to find the law. It's another kind of miracle. You see, knowing a law to figure out that such and such is going to do something, and then have nature do it — OK, that's pretty good. But to look at other aspects and to guess, and to know that there's a pattern under there, and to tell nature that in this experiment she's going to do that — not by deduction, strictly speaking, from what's known but by guessing from what's known — it seems a wonderful thing to me. And I always wanted to do that.

– Interview with Charles Weiner, June 28, 1966 (Niels Bohr Library and Archives with the Center for the History of Physics)

It was because it was a challenge, for its own sake — completely, always, for its own sake. It was the excitement, the fun, of working this thing out.

– Interview with Charles Weiner, March 4, 1966 (Niels Bohr Library and Archives with the Center for the History of Physics)

Even in my crazy book, I didn't emphasize — but it is true — that I worked as hard as I could at drawing, at deciphering Mayan, at drumming, at cracking safes, etc. The real fun of life is this perpetual testing to realize how far out you can go with any potentialities.

– Letter to Mr. V. A. Van Der Hyde, July 1986 (*Perfectly Reasonable Deviations from the Beaten Track*, p. 414)

The way I think of what we're doing is, we're exploring — we're trying to find out as much as we can about the world. People say to me, "Are you looking for the ultimate laws of physics?" No, I'm not. I'm just looking to find out more about the world. If it turns out there is a simple ultimate law which explains everything, so be it; that would be very nice to discover. If it turns out it's like an onion, with millions of layers, and we're sick and tired of looking at the layers, then that's the way it is. But whatever it comes out, it's nature, and she's going to come out the way she is! Therefore when we go investigate it we shouldn't predecide what it is we're going to find, except to find out more.

– BBC, "The Pleasure of Finding Things Out," 1981

I often think about that, especially when I'm teaching some esoteric technique such as integrating Bessel functions. When I see equations, I see the letters in colors — I don't know why. As I'm talking, I see vague picture of Bessel functions from Jahnke and Emde's book, with light-tan js, slightly violet-bluish ns, and dark brown xs flying around. And I wonder what the hell it must look like to the students.

– *What Do You Care What Other People Think?*, p. 59

My interest in science is to simply find out more about the world, and the more I find out, the better it is. I like to find out.

– BBC, "The Pleasure of Finding Things Out," 1981

I'll never make that mistake again, reading the experts' opinions. Of course, you only live one life: You make all your mistakes, and learn what not to do — and that's the end of you.

– *Surely You're Joking, Mr. Feynman!*, p. 255

Since then I never pay any attention to anything by "experts." I calculate everything myself.

– *Surely You're Joking, Mr. Feynman!*, p. 255

It cannot interest you until you deeply understand the problem and its intricacies. Then every subject is interesting.

– Notes

The work is not done for the sake of an application. It is done for the excitement of what is found out.

– "The Uncertainty of Science," John Danz Lecture Series, 1963

Learn what the rest of the world is like. The variety is worthwhile.

– *Surely You're Joking, Mr. Feynman!*, p. 63

The next day at the meeting, I saw [Murray] Slotnick and said, "Slotnick, I worked it out last night, I wanted to see if I got the same answers you do. I got a different answer for each coupling — but, I would like to check in detail with you because I want to make sure of my methods." And, he said, "what do you mean you worked it out last night, it took me six months!" That was a thrilling moment for me, like receiving the Nobel Prize, because that convinced me, at last, I did have some kind of method and technique and understood how to do something that other people did not know how to do. That was my moment of triumph in which I realized I really had succeeded in working out something worthwhile.

– From *Nobel Lectures, Physics 1963–1970*, Elsevier Publishing
Company, Amsterdam, 1972

It has to do with curiosity. It has to do with people wondering what makes something do something. And then to discover that if you try to get answers, that they're related to each other. The things that make the wind make the waves, and the motion of water is like the motion of air is like the motion of sand. The fact that things have common features, turns out more and more universal. What we're looking for is how everything works. What makes everything work.

– Yorkshire Television interview, "Take the World from Another Point of View," 1972

It's curiosity, as to where we are, what we are. It's very much more exciting to discover that we're on a ball, half of us sticking upside down, it's spinning around in space, there's a mysterious force which holds us on, going around a great big glob of gas that's burning, fueled by a fire that's completely different from any fire we can make (well, now we can make that fire — nuclear fire), but that's a much more exciting story to many people than the tales which other people used to make up, who worried about the universe; that we were living on the back of a turtle, or something like that. They were wonderful stories, but the truth is so much more remarkable. The pleasure of physics to me, is that it's revealed that the truth is so remarkable, so amazing.

– Yorkshire Television interview, "Take the World from Another Point of View," 1972

The same thrill, the same awe and mystery, come again and again when we look at any problem deeply enough. With more knowledge comes deeper, more wonderful mystery, luring one on to penetrate deeper still. Never concerned that the answer may prove disappointing, but with pleasure and confidence we turn over each new stone to find unimagined strangeness leading on to more wonderful questions and mysteries — certainly a grand adventure!

– "The Value of Science," December 1955

[On physics problems:] Most of the time I don't solve them. Once in a great while I do. And since the problems I've chosen to work on are rather big, hard problems that nobody else has solved, when you solve something nobody else has solved, you've got a little pride. You get a kick out of doing it.

– "The Remarkable Dr. Feynman," *Los Angeles Times Magazine*, April 20, 1986

I don't know anything, but I do know that everything is interesting if you go into it deeply enough.

– *Omni* interview, February 1979

The only thing in the world is not applications. It's interesting in understanding what the world is made of. It's the same interest, the curiosity of man that makes him build telescopes. What is the use of discovering the age of the universe? Or what are these quasars that are exploding at long distances? I mean what's the use of all that astronomy? There isn't any. Nonetheless, it's interesting.

– Future for Science interview

A guy came into my room and found me leaning out of a wide-open window in the dead of winter, holding a pot in one hand and stirring with the other. I was curious as to whether jello would coagulate if you kept it moving all the time.

– *What Do You Care What Other People Think?*, p. 56

I'm not answering your question, but I'm telling you how difficult a "why" question is. You have to know what it is that you're permitted to understand and allow to be understood and know, and what it is you're not.

– BBC, "Fun to Imagine" television series, 1983

If you heat a rubber band, it will pull more strongly. For instance, if you hang a weight with a rubber band, and put a match to it, it's kind of fun to watch it rise.

– BBC, "Fun to Imagine" television series, 1983

I love puzzles: One guy tries to make something to keep another guy out; there must be a way to beat it.

– *Surely You're Joking, Mr. Feynman!*, p. 139

You should never be afraid that something new will come out. It will come out in due time, and then you'll try to understand it. Of course, it'll be very exciting!

– Audio recording of Feynman Lectures on Physics, Lecture 3, October 3, 1961, Q&A

It's the way I study — to understand something by trying to work it out or, in other words, to understand something by creating it. Not creating it one hundred percent, of course, but taking a hint as to which direction to go but not remembering the details. Those you work out for yourself.

– *Feynman Lectures on Computation*, p. 15

Isn't interesting to live in our time and have such wonderful puzzles to work on?

– Letter to Dr. Victor F. Weisskopf, January–February 1961

He often started to talk about things like this: "Suppose a man from Mars were to come down and look at the world." It's a very good way to look at the world.

– National Science Teachers Association Fourteenth Convention lecture, "What Is Science?" April 1966

I kind of try to imagine what would have happened to me if I'd lived in today's era. I'm rather horrified. I think there are too many books that the mind gets boggled. If I got interested, I would have so many things to look at, I would go crazy. It's too easy. Maybe. Maybe not. Maybe this is just an old fashioned point of view.

– Interview with Charles Weiner, March 4, 1966 (Niels Bohr Library and Archives with the Center for the History of Physics)

And sometimes people say, "How is it you're suddenly working on this?" It's just that I finally got some success. I work often on a large range of things that don't work out. Then there's silence.

And then people say, "Why are you suddenly doing this?" Well, yeah, I finally got somewhere on this. It's not that I suddenly did it.

– Interview with Charles Weiner, June 28, 1966 (Niels Bohr Library and Archives with the Center for the History of Physics)

What we're looking for is how everything works and what makes everything work. And what happens first in history is that we discover first the things that are on the face of it that are obvious, and gradually we ask more questions and dig in a little deeper to things we need to do a little more complicated experiment to find out.

– Interview with Yorkshire Television program, "Take the World from Another Point of View," 1972

Many of the other people who tell you about Los Alamos know somebody up in some higher echelon of governmental organization or something, worried about some big decision. I worried about no big decisions. I was always flittering about underneath somewhere.

– UCSB talk, "Los Alamos from Below," February 1975

Maybe the reason we make so much progress in physics is because it's so easy.

– *Great American Scientists*, p. 24

I studied metallography, for instance, because it was a field I didn't know anything about. I was always interested to learn something about which I didn't know anything — to see what would happen, you know, in metallurgy, and metallography. I remember that course in particular. That's when I discovered for the first time the very great use of your knowledge of physics, the universality.

– Interview with Charles Weiner, March 5, 1966 (Niels Bohr
   Library and Archives with the Center for the History of Physics)

I think the right way, of course, is to say that what we have to look at is the whole structural interconnection of the thing; and that all the sciences, and not just the sciences but all the efforts of intellectual kinds, are an endeavor to see the connections of the hierarchies, to connect beauty to history, to connect history to man's psychology, man's psychology to the working of the brain, the brain to the neural impulse, the neural impulse to the chemistry, and so forth, up and down, both ways.

– *The Character of Physical Law*, p. 125

I can't answer adult questions. They're bad questions. Usually they want to know the meaning of some new word they've seen and it's something they'll never understand. I hate adults. Younger people are curious about nature.

– *Columbia Dispatch*, October 22, 1966

I decided that, when I was a kid, I used to enjoy the subject for the fun of it. I used to like nature and do it for the fun of it. So what I ought to do is play games with it, just whatever was curious and interesting to me — I should just play. You see? Just like I was when I was a kid — try to find a relation between things, do this, do that, whatever I felt like. I don't have to do this problem because it's important or that problem because it's important, or everybody expects me to do something.

– Interview with Charles Weiner, June 27, 1966 (Niels Bohr Library and Archives with the Center for the History of Physics)

I don't really read a lot of what the other guy does. I read what his assumptions are and if they seem reasonable then I work out the conclusions. I don't need to read how he works out the conclusions most of the time.

– Interview with Charles Weiner, June 28, 1966 (Niels Bohr Library and Archives with the Center for the History of Physics)

I don't wanna take this stuff seriously. I think we should just have fun imagining it and not worry about it. There's no teacher going to ask you questions at the end. Otherwise, it's a horrible subject.

– BBC, "Fun to Imagine" television series, 1983, regarding particle physics

Sometimes the truth is discovered first, and the beauty or "necessity" of that truth seen only later.

– "Structure of the Proton," Niels Bohr Medal lecture given in Copenhagen, Denmark, October 1973

But if you keep proving stuff that others have done, getting confidence, increasing the complexities of your solutions — for the fun of it — then one day you'll turn around and discover that nobody actually did that one!

– *Feynman Lectures on Computation*, p. 16

I want to do scientific research — that is, to find out more about how the world works. And that is not a secret; that work is not secret.

– Interview for Viewpoint

The hardest part of the helium problem was done by physical reasoning alone, without being able to write anything. By just standing — you know, I remember kind of leaning against the kitchen sink, you see, and looking at it, and just thinking. And it was very, very interesting to be able to push through that doggoned thing without having stuff to write.

– Interview with Charles Weiner, June 28, 1966 (Niels Bohr Library and Archives with the Center for the History of Physics)

When you explain a "why," you have to be in some framework in which you've allowed something to be true. Otherwise, you're perpetually asking why.

– BBC, "Fun to Imagine" television series, 1983

You begin to get a very interesting understanding of the world and all its complications. If you try to follow anything up, you go deeper and deeper in various directions.

– BBC, "Fun to Imagine" television series, 1983

All we know, is that as we go along we find we can amalgamate pieces, and then find pieces that don't fit, and keep trying to put the jigsaw puzzle together. But whether there's a finite number of pieces, and whether there's a border to this jigsaw puzzle or not, is of course, not known. And it will never be known until we — if we do — ever finish the picture.

 – Audio recording of Feynman Lectures on Physics, Lecture 2,
    September 29, 1961

That was the beginning, and the idea seemed so obvious to me and so elegant that I fell deeply in love with it. And, like falling in love with a woman, it is only possible if you do not know much about her, so you cannot see her faults. The faults will become apparent later, but after the love is strong enough to hold you to her. So, I was held to this theory, in spite of all difficulties, by my youthful enthusiasm.

 – From *Nobel Lectures, Physics 1963–1970*, Elsevier Publishing
    Company, Amsterdam, 1972

We live in a heroic age. We live in a moment that'll never come again. These discoveries cannot be made twice. One doesn't discover America two or three times in succession, really, and one doesn't discover the laws of nuclear forces or electricity more than once.

 – BBC, "Strangeness Minus Three," 1964

That's the way it always is in these pinnacle discoveries. The big pile-up of stuff, all the old things that you've thought of before you try again and again, but the great discovery always involves a great philosophical surprise.

– BBC, "Strangeness Minus Three," 1964

The chance is high that the truth lies in the fashionable direction. But, on the off-chance that it is in another direction — a direction obvious from an unfashionable view of field theory — who will find it? Only someone who has sacrificed himself by teaching himself quantum electrodynamics from a peculiar and unfashionable point of view; one that he may have to invent for himself.

– From *Nobel Lectures, Physics 1963–1970*, Elsevier Publishing Company, Amsterdam, 1972

How Physicists Think

There is always the possibility of proving any definite theory wrong;
but notice that we can never prove it right. Suppose that you invent
a good guess, calculate the consequences, and discover every time
that the consequences you have calculated agree with experiment.
The theory is then right? No, it is simply not proved wrong.

– *The Character of Physical Law*, p. 157

People are still designing new gyros, new devices, new ways, and it may well be that one of them will solve the problems, for instance, this inanity of having to have the axle bearings so accurate. If you play with the gyro for a while you will see that the friction on its axle is not small. The reason is, if the bearings were made too frictionless, the axle would wobble, and you'd have to worry about that tenth of a millionth of an inch — which is ridiculous. There must be a better way.

– *Feynman's Tips on Physics*, p. 129

It is important to realize that in physics today, we do not have any knowledge of what energy is.

– Audio recording of Feynman Lectures on Physics, Lecture 4, October 6, 1961

Thinking is nothing but talking to yourself inside.

– *What Do You Care What Other People Think?*, p. 54

Physicists like to think that now all you have to do is say, "These are the conditions, now what happens next?"

– *The Character of Physical Law*, p. 114

Everything I would do like this, I would always have a practical problem to exemplify it. I have always thought the thing was no good unless you could use it somehow.

– Interview with Charles Weiner, March 4, 1966 (Niels Bohr Library and Archives with the Center for the History of Physics)

A great deal of formulation work is done in writing the paper, organizational work, organization. I think of a better way, a better way, a better way of getting there, of proving it. I never do much — I mean, it's just cleaner, cleaner and cleaner. It's like polishing a rough-cut vase. The shape, you know what you want and you know what it is. It's just polishing it. Get it shined, get it clean, and everything else.

– Interview with Charles Weiner, March 4, 1966 (Niels Bohr Library and Archives with the Center for the History of Physics)

When I think about something, I go along in a certain way, and then I get balled up, and then I go back, and I think — I get mixed up easily. I easily get into confusion, which is the horror of the whole business when you're thinking. It's like building a pile of cards and the whole thing collapses, and you keep going and it collapses.

– Interview with Charles Weiner, June 28, 1966 (Niels Bohr Library and Archives with the Center for the History of Physics)

I try always to work with the least knowledge possible, with a little knowledge of what other people are doing, because I feel more happy being more individual, if I am not following the line and getting confused by what they say.

— Interview with Charles Weiner, June 28, 1966 (Niels Bohr Library and Archives with the Center for the History of Physics)

I don't like to judge other people, or their work, at all. I don't. I don't want to judge somebody else's work.

— Interview with Charles Weiner, June 28, 1966 (Niels Bohr Library and Archives with the Center for the History of Physics)

I'd been traveling every summer somewhere, and I thought this time, where am I going to go? And I said, the hell with it, I don't feel like traveling. Instead of traveling, I'll do biology experiments. I'll go into a different field, instead of going into a different country.

— Interview with Charles Weiner, June 28, 1966 (Niels Bohr Library and Archives with the Center for the History of Physics)

I had a principle that everything that I wrote, I should understand inside out; that there was just a little bit less written than what I knew; and that whatever I wrote would be right. I didn't like the papers that somebody would write; suggesting an idea which in three months they find is cockeyed.

— Interview with Charles Weiner, June 28, 1966 (Niels Bohr Library and Archives with the Center for the History of Physics)

It was to me a rather annoying thing, because I realized what it would mean is all this noise and all this trouble and wild business, you know? There'd be newspapermen for whom I had no respect, publicity for which I had no respect. This world is so full of hot air, and just extra propaganda junk today, that it's not real. I just don't want to get involved with all that stuff, and I didn't know how I was going to escape it. I still imagined that I could by not answering the phone. So I took the telephone off the hook.

– Interview with Charles Weiner, June 28, 1966 (Niels Bohr Library and Archives with the Center for the History of Physics)

It has turned out that many things that bothered me that I thought that I didn't understand because I didn't know enough about the subject, turned out that I didn't understand them because they weren't logical, they weren't valid.

– Interview with Charles Weiner, February 4, 1973 (Niels Bohr Library and Archives with the Center for the History of Physics)

But I don't work in a straight line. So I leap ahead in one direction and I'm expecting to close up with the other and make a finished product, OK? But the leap that went in the forward direction works. I think that's not all there is to it. I think I'll be able to make a bigger do, and I work on it, but it doesn't get any bigger. I get confused or something. But the leap ahead was very important.

– Interview with Charles Weiner, February 4, 1973 (Niels Bohr Library and Archives with the Center for the History of Physics)

You know how scientists, they usually look at the rest of the world, think of how absurd everything is, how stupid everybody is, how dopey the Navy guys are that measured the distances different, so ha-ha, but you should be ashamed of yourself, we should be ashamed of ourselves, for there's no difference, for example, in the ways we measure anything.

– Audio recording of lecture on relativity, Douglas Advanced Research Laboratory, 1967

If you're ever running for an airplane in a taxi, and you suddenly discover that you're using Daylight Saving Time instead of Standard Time or something like that, and you don't know whether you're going to make the plane or not, and you try to figure out, which way? You know how complicated it is.

– Audio recording of lecture on relativity, Douglas Advanced Research Laboratory, 1967

I took this stuff that I got out of your seal and put it in ice water, and I discovered that when you put some pressure on it for a while and then undo it it doesn't stretch back. It stays the same dimension. In other words, for a few seconds at least and more seconds than that, there is no resilience in this particular material when it is at a temperature of 32 degrees.

– Transcript of the hearing of the Presidential Commission on the Space Shuttle Accident, February 11, 1986

The thing that doesn't fit is the thing that's the most interesting, the part that doesn't go according to what you expected.

– BBC, "The Pleasure of Finding Things Out," 1981

The physicists should be ashamed of themselves: Astronomers keep asking, "Why don't you figure out for us what will happen if you have a big mass of junk pulled together by gravity and spinning? Can you understand the shapes of these nebulae?" And nobody ever answers them.

– *Feynman's Tips on Physics*, p. 127

I believe that has some significance for our problem.

– Transcript of the hearing of the Presidential Commission on the Space Shuttle Accident, February 11, 1986

Once I got into physics I forgot who I was talking to.

– Interview with Charles Weiner, June 28, 1966 (Niels Bohr Library and Archives with the Center for the History of Physics)

You will have to brace yourselves for this — not because it is difficult to understand, but because it is absolutely ridiculous: All we do is draw little arrows on a piece of paper — that's all!

– *QED: The Strange Theory of Light and Matter*, p. 24

The paradox is only a conflict between reality and your feeling of what reality ought to be.

– *Feynman Lectures on Physics*, vol. 3, pp. 18–19

Do you think that it is not a paradox, but that it is still very peculiar? On that we can all agree. It is what makes physics fascinating.

– *Feynman Lectures on Physics*, vol. 3, pp. 18–19

The real problem in speech is not precise language. The problem is clear language. The desire is to have the idea clearly communicated to the other person. It is only necessary to be precise when there is some doubt as to the meaning of a phrase, and then the precision should be put in the place where the doubt exists. It is really quite impossible to say anything with absolute precision, unless that thing is so abstracted from the real world as to not represent any real thing.

– "New Textbooks for the 'New' Mathematics," *Engineering and Science* 28, no. 6 (March 1965)

And you can re-create the things that you've forgotten perpetually — if you don't forget too much, and if you know enough. In other words, there comes a time where you'll know so many things that as you forget them, you can reconstruct them from the pieces that you can still remember. It is therefore of first-rate importance that you know how to triangulate — that is, to know how to figure something out from what you already know. It is absolutely necessary.

– *Feynman's Tips on Physics*, p. 39

There's a unit of time I must use here if I use miles here. *c* equals the speed of light. If you wish, that formula tells you what I just said. Which was you convert time into miles by asking, what's the unit of time? The time to go one mile. You want to measure this in kilometers? Okay, and the unit of time is the time to go one kilometer. And if you use this interval to measure time, instead of seconds, then the formula's right.

– Audio recording of lecture on relativity, Douglas Advanced Research Laboratory, 1967

Incidentally, from the point of view of basic physics, the most interesting phenomena are of course, the new places, the places that don't work, not the places that do work. Because there's where we'll discover new rules.

– Audio recording of Feynman Lectures on Physics, Lecture 2, September 29, 1961

It will not do to memorize the formulas, and to say to yourself, "I know all the formulas; all I gotta do is figure out how to put 'em in the problem!"

– *Feynman's Tips on Physics*, p. 38

First, I'm going to tell you what the theory is; I'll tell you what it looks like, what we do to make the calculations, just what the thing is because otherwise how are you going to understand what "world picture" this thing is? And it is a "world picture" because it describes all the phenomena (except radioactivity and gravity) in the world! That's a lot of phenomena! It should explain, if everything is thoroughly understood, the laughter of the audience when you make a dummy remark!

– "QED: Photons — Corpuscles of Light," Sir Douglas Robb Lectures, University of Auckland, June 1979

In fact, the total amount that a physicist knows is very little. He has only to remember the rules to get him from one place to another and he is all right.

– *The Character of Physical Law*, p. 45

I learned that really physics is a very useful background for
what looks like different fields; that the world is the same, the
physical laws are not so un-useful, you know what I mean? They
work. Yeah, they work, and you can use the ideas in different
fields, and you are ahead of the other guys, because there are
a large number of things that are self-evident to you that they
have to learn. But of course, you have to learn by experience
too. I'm not trying to say just — both together are much better
than anybody. But it is true that studying physics is good all
over the place.

– Interview with Charles Weiner, March 5, 1966 (Niels Bohr
Library and Archives with the Center for the History of Physics)

It's a very hard job. It's lots of work. So what do we do it for?
Because of the excitement, because of the fact that each time we
get one of these things, we have a terrific — El Dorado — we have
a new view of nature. We see the ingenuity, if I may put it that way,
of nature herself. The peculiarity of the way she works. It takes a
terrible strain on the mind to understand these things. And the
real value of the development of the science, in this connection, the
thing that makes me go on, is this, the difficulty of understanding
it. That these apes, stand around looking at nature and find that to
really catch on, they have to polish their minds to the last.

– BBC, "Strangeness Minus Three," 1964

I happen to know this, and I happen to know that, and maybe I
know that; and I work everything out from there. Tomorrow I may

forget that this is true, but remember that something else is true, so I can reconstruct it all again. I am never quite sure of where I am supposed to begin or where I am supposed to end. I just remember enough all the time so that as the memory fades and some of the pieces fall out I can put the thing back together again every day.

– *The Character of Physical Law*, p. 47

The Quantum World

We physicists are always checking to see if there is something the matter with the theory. That's the game, because if there is something the matter, it's interesting! But so far, we have found nothing wrong with the theory of quantum electrodynamics. It is therefore, I would say, the jewel of physics — our proudest possession.

– *QED: The Strange Theory of Light and Matter*, p. 8

What I am going to tell you about is what we teach our physics students in the third or fourth year of graduate school — and you think I'm going to explain it to you so you can understand it? No, you're not going to be able to understand it. Why, then, am I going to bother you with all this? Why are you going to sit here all this time, when you won't be able to understand what I am going to say? It is my task to convince you not to turn away because you don't understand it. You see, my physics students don't understand it. That is because I don't understand it. Nobody does.

– *QED: The Strange Theory of Light and Matter*, p. 9

The science fiction writers who have interpreted my view of the positron as an electron going backward in time have not realized that the theory is completely consistent with causality principles and in no way implies that we can travel backward in time.

– Correspondence with David Paterson (BBC), February 1976

All my mature life I have been trying to distill the strangeness of quantum mechanics into simpler and simpler circumstances. I have given many lectures of ever increasing simplicity and purity.

– Letter to Dr. N. David Mermin, March 1984 (*Perfectly Reasonable Deviations from the Beaten Track*, p. 368)

Maybe gravity is a way that quantum mechanics fails at large distances.

– Letter to Dr. Victor F. Weisskopf, January–February 1961

Before and after are not absolute ideas; they depend on the point of view. It is similar to the question of what is in front and what is in back. If I turn a little bit, I can change the arrangement. Two things may appear to be at the same distance from one man, but appear to be at different distances from another. Likewise, it is true that two events which may appear to be at the same from one point of view may not do so from another. This leads us to the idea of the representation of time as a fourth geometrical dimension.

– From notes for "About Time" program, 1957

I really do believe that quantum mechanics is fundamentally correct, and that all this is simply psychological trouble. It is extremely difficult to get used to it because it's so much common sense and common knowledge that gets this idea that when you're not looking at something, it's either this way or that way. And to be able to say, "By God, you can't even say it's either this way or that way when you don't look at it? But it must be either" No, you can't say that, or you're going to get in trouble! They say that can't be so bad. It must be that Nature isn't quite like that.

– Esalen lecture, "Quantum Mechanical View of Reality (Part 1)," October 1984

Now I come to the greatest shame of theoretical physics. For the last twenty-five years, well it's more than that, it's nearly forty years, field theories have been in existence, and still no one can compute most of their consequences exactly. Even the predictions of Yukawa's field theory, for example, can't be calculated exactly. We can calculate approximately in electrodynamics only because the coupling is small. We make a series expansion in the coupling constant. When we can't make the series expansion we're too stupid to figure out what the consequences are. That's a sin. It's one of the reasons we are not making much progress.

– Caltech lecture on particles, 1973

One is presumptuous if one says, "We're going to find the ultimate particle, or the unified field laws," or "the" anything. If it turns out surprising, the scientist is even more delighted. You think he's going to say, "Oh, it's not like I expected, there's no ultimate particle, I don't want to explore it?" No, he's going to say, "What the hell is it, then?"

– *Omni* interview, February 1979

The behavior of things on a small scale by quantum mechanics is weird and wonderful and theoretical physicists have always tried to delight the general public with the wonders of the small world.

– "Theory and Applications of Mercereau's Superconducting Circuits," October 1964

The most interesting problems today — and certainly the most practical problems — are obviously in solid-state physics. But someone said there is nothing so practical as a good theory, and the theory of quantum electrodynamics is definitely a good theory!

– *QED: The Strange Theory of Light and Matter*, p. 114

And the best thing to do is to relax and enjoy this: the tiny-ness of us, and the enormity of the rest of the universe. Of course, if you're feeling depressed by that, you can always look at it the other way and think of how big you are compared to the atoms and the parts of atoms, and then you're an enormous universe to those atoms, and then you can sort of stand in the middle and enjoy everything in both ways.

– BBC, "Fun to Imagine" television series, 1983

We may think of two quarks and antiquarks with a sausage around them — but to move the sausage involves field inertia or effective mass proportional to the length of the sausage.

– "Mass Varying with Position," *Physics* 230, 1987 (R. P. Feynman Papers, California Institute of Technology Archives)

The principles of physics, as far as I can see do not speak against the possibility of maneuvering things atom by atom. It is not an attempt to violate any law; it is something in principle that can be done; but in practice, it has not been done because we are too big.

"There's Plenty of Room at the Bottom," December 1959

It is very important to know that light behaves like particles, especially for those of you who have gone to school, where you were probably told something about light behaving like waves. I'm telling you the way it does behave — like particles.

– *QED: The Strange Theory of Light and Matter*, p. 15

There was a time when the newspapers said that only twelve men understood the theory of relativity. I do not believe there ever was such a time. There might have been a time when only one man did, because he was the only guy who caught on, before he wrote his paper. But after people read the paper a lot of people understood the theory of relativity in some way or other, certainly more than twelve. On the other hand, I think I can safely say that nobody understands quantum mechanics.

– *The Character of Physical Law*, p. 129

The theory of quantum electrodynamics describes nature as absurd from the point of view of common sense. And it agrees fully with experiment. So I hope you can accept nature as she is — absurd.

– *QED: The Strange Theory of Light and Matter*, p. 10

Our first experiment to try to understand the proton as a complicated object, like a watch, was to knock two of these watches together at high energy and to look at what kind of gear wheels and so forth came out at what angles. This is the typical hadron-hadron collision, and it involves two unknowns, both the target and the missile.

– Oersted Medal acceptance speech, 1972

We have often made great advances at physics by recognizing that the complexity of things at one level is the result of the fact that these things are composed of simpler elements at another level.

— "Structure of the Proton," Niels Bohr Medal lecture given in Copenhagen, Denmark, October 1973

[On the existence of quarks:] There are, however, a number of theoretical arguments against this idea. So strong are these arguments that at first they seemed to lead to paradoxes. But one by one we were learning how it may be possible to get around these paradoxes. We are perhaps getting the first glimpses of a truly dynamical theory of the hadrons.

— "Structure of the Proton," Niels Bohr Medal lecture given in Copenhagen, Denmark, October 1973

If the experiments continue to confirm the need for quarks in protons then this is the way the theory will apparently develop: Quarks of three colors, so nine in all. And eight kinds of gluons. This part sounds elaborate but is mathematically simple. And a long-range force — which sounds simple but appears mathematically a bit unnatural. Suggestions to explain this long-range force, such as Kauffmann's, all seem a little awkward and not to have an inner beauty we usually expect from truth. But sometimes the truth is discovered first and the beauty or "necessity" of that truth seen only later.

— "Structure of the Proton," Niels Bohr Medal lecture given in Copenhagen, Denmark, October 1973

Beside our eight gluons and nine quarks there would still be the electron, muon, photon, graviton, and two neutrinos, so we would still leave a new proliferation of particles to be analyzed by the next generation. Will they find them all composed of yet simpler elements at yet another level?

– "Structure of the Proton," Niels Bohr Medal lecture given in Copenhagen, Denmark, October 1973

The proton and neutron are only two out of about four hundred known varieties of things. This is a terrible mess, worse than chemistry in Mendeleev's time by about a factor of four or five!

– Caltech lecture on particles, 1973

If the predictions keep on working, then you can keep saying, "Yes, things are not really made out of partons, but they act as if they are made out of partons in this respect, and in that respect, and in that respect." That's what reality is in physics, an idea which fits an even wider class of experiments. In the days when people said things are made out of atoms, there were objections that you could get the same results from thermodynamics. Only the thermodynamics properties were right. In that particular case, as experiments went on and on, the proposition that things were made of atoms turned out to be of a much greater generality and correctness than just thermodynamics.

– Caltech lecture on particles, 1973

There has never been a satisfactory model of reflection of light from thin surfaces or, after that, from any other phenomenon.

Satisfactory in their old fashioned, classical view: a logical hocus-pocus has to be done quantum mechanically in order to describe these things.

– Esalen lecture, "Quantum Mechanical View of Reality (Part 2)," October 1984

The behavior of sub-nuclear systems is so strange compared to the ones the brain evolved to deal with that the analysis has to be very abstract: to understand ice, you have to understand things that are themselves very unlike ice.

– *Omni* interview, February 1979

If they'd explain that this is their best guess but so few of them do; instead, they seize on the possibility that there may not be any ultimate fundamental particle, and say that you should stop work and ponder with great profundity. "You haven't thought deeply enough, first let me define the world for you." Well, I'm going to investigate it without defining it!

– *Omni* interview, February 1979

I've entertained myself always by squeezing the difficulty of quantum mechanics into a smaller and smaller place.

– MIT conference, May 1981

The incompleteness of our present view of quantum electrodynamics should not blind us to the enormous progress that has been made.

– "The Present Situation in Quantum Electrodynamics," Solvay conference, 1961

In turns out that on the microscopic scale, all the laws of physics are exactly reversible: forwards in time, backwards in time — looks the same. But all those phenomena (and there are many of course: life and frying eggs are two examples) that go in one direction only have to be interpreted by the complexity of circumstances, that there are so many particles getting mixed up.

– "QED: Electrons and Their Interactions," Sir Douglas Robb
   Lectures, University of Auckland, June 1979

So what I'm going to talk about in this lecture is all of physics, really — all of known physics at the moment, plus a lot that's guessed. So the lecture's going to be even more extensive than the ones before because there I was dealing with only two particles, and now I have to deal with two dozen particles!

– "QED: Electrons and Their Interactions," Sir Douglas Robb
   Lectures, University of Auckland, June 1979

It's lucky we have such a large-scale view of everything that we can see them as things without having to worry about all these little atoms all the time.

– BBC, "Fun to Imagine" television series, 1983

When we were talking about the atoms, one of the troubles that people had with the atoms was that they're so tiny, and it's so hard to imagine the scale. That the size of what the atoms are, compared to an apple, is the same scale as the size of the apple compared to the earth. And that's kind of a hard thing to take. You have to go through these things all the time, and people find these numbers inconceivable. And I do, too!

– BBC, "Fun to Imagine" television series, 1983

[On atoms:] Now, what you do is, you just change your scale. You're just thinking of small balls; you don't try to think of exactly how small they are too often, or you get kind of a bit nutty.

– BBC, "Fun to Imagine" television series, 1983

It was about the beginning of the 1900s that it was discovered that light, as a matter of fact, behaves like particles, which was a terrible shock after great success with the wave theory. And then the problem of trying to see the how particles could make these wave-like phenomena so easily explained by waves became known as the "wave–particle duality." Light behaves like particles on Thursdays and like waves on Tuesdays, and that, of course, is not a satisfactory theory.

– "QED: Fits of Reflection and Transmission," Sir Douglas Robb Lectures, University of Auckland, 1979

I remember being so distinctly impressed by the boldness and novelty of using quantum mechanics to describe such macroscopic things.

– Letter to Dr. Scully, February 1974

If we will only allow that, as we progress, we remain unsure, we will leave opportunities for alternatives. We will not become enthusiastic for the fact, the knowledge, the absolute truth of the day, but remain always uncertain. In order to make progress, one must leave the door to the unknown ajar.

– Galileo Symposium, "What Is and What Should Be the Role of Scientific Culture in Modern Society," September 1964

But the thing that's unusual about good scientists is that they're not so sure of themselves as others usually are. They can live with steady doubt, think "maybe it's so" and act on that, all the time knowing it's only "maybe."

– *Omni* interview, February 1979

Multiply the importance of a problem by your ability to do something about it.

– Notes

The behavior of things in the small scale is so fantastic! It is so wonderfully different! So marvelously different than anything that behaves on a large scale.

– BBC, "Fun to Imagine" television series, 1983

When these new ideas of relativity and quantum mechanics were developed early in this century they appeared so strange that many conservative people hoped they would ultimately be proven wrong, but the last half-century of experiment of ever greater energy, scope and accuracy, have only continued to confirm them.

– "Structure of the Proton," Niels Bohr Medal lecture given in Copenhagen, Denmark, October 1973

I have tried to explain that all the improvements of relativistic theory were at first more or less straightforward, semi-empirical shenanigans. Each time I would discover something, however, I would go back and I would check it so many ways, compare it to every problem that had been done previously in electrodynamics (and later, in weak coupling meson theory) to see if it would always agree, and so on, until I was absolutely convinced of the truth of the various rules and regulations which I concocted to simplify all the work.

- From *Nobel Lectures, Physics 1963–1970*, Elsevier Publishing Company, Amsterdam, 1972

What happens inside will be the same whether that thing is standing still or is coasting along at a uniform speed.

- Audio recording of lecture on relativity, Douglas Advanced Research Laboratory, 1967

I never objected to what other people would immediately have objected to, you know. All the books would say we can't use advanced waves because this would mean effects would precede causes. But things like that never bothered me. I don't give a darn. I never thought in terms of cause and effect necessarily, anything.

- Interview with Charles Weiner, March 5, 1966 (Niels Bohr Library and Archives with the Center for the History of Physics)

[On quantum mechanics:] We can't pretend to understand it since it affronts all our commonsense notions. The best we can do is to describe what happens in mathematics, in equations and that's very difficult. What is even harder is trying to decide what the equations mean. That's the hardest thing of all, but then that is what makes this business of the physics of the nucleus so exciting. That's why I do it. No one can honestly say this is going to lead to practical knowledge. We can't say we're going to get a new source of energy. We do it because it's what we're good at and because it is a great adventure in imagination.

– BBC, "Horizon: The Hunting of the Quark," May 1974

It used to be thought that atoms were small, and that was a limit of measurement, but at the present time, with the new instruments and design during all this time, we've been able to make instruments that can test this theory. Now the distances can be described this way: If the atom is made one hundred kilometers on a side, then we're measuring with one-centimeter accuracy.

– "QED: Photons — Corpuscles of Light," Sir Douglas Robb Lectures, University of Auckland, June 1979

Doing these particle experiments is easy — it's far more difficult to analyze the results what they mean. What could you make of a collision in which two mini cars crashed head on at high speed and produced a Rolls Royce — or two Rolls Royces crashing to make a motor cycle?

– BBC, "Horizon: The Hunting of the Quark," May 1974

Science and Society

You cannot understand science and its relation to anything unless you understand and appreciate the great adventure of our time. You do not live in your time unless you understand that this is a tremendous adventure and a wild and exciting thing.

– "The Uncertainty of Science," John Danz Lecture Series, 1963

It is necessary to teach both to accept and to reject the past with a kind of balance that takes considerable skill.

– National Science Teachers Association Fourteenth Convention lecture, "What Is Science?" April 1966

The conclusion from all the researchers is that all people in the world are as dopey as can be and the only way to tell them anything is to perpetually insult their intelligence.

– "The Unscientific Age," John Danz Lecture Series, 1963

In any decision for action, when you have to make up your mind what to do, there is always a "should" involved, and this cannot be worked out from "If I do this, what will happen?" alone. You say, "Sure, you see what will happen, and then you decide whether you want it to happen or not." But that last stop, the decision whether you want it to happen or not, is the step the scientists cannot help you with.

– "The Uncertainty of Science," John Danz Lecture Series, 1963

I think that to say these are scientific problems is an exaggeration. They are far more humanitarian problems. The fact that how to work the power is clear, but how to control it is not, is something not so scientific and is not something that the scientist knows so much about.

– "The Uncertainty of Science," John Danz Lecture Series, 1963

I have a large number of uncomfortable feelings about the world.

– "The Unscientific Age," John Danz Lecture Series, 1963

The attitude of the populace is to try to find the answer instead of trying to find a man who has a way of getting at the answer.

– "The Unscientific Age," John Danz Lecture Series, 1963

To be prejudiced against mind reading a million to one does not mean that you can never be convinced that a man is a mind reader.

– "The Unscientific Age," John Danz Lecture Series, 1963

I would like to point out that people are not honest. Scientists are not honest at all, either. It's useless. Nobody's honest. Scientists are not honest. And people usually believe that they are. That makes it worse. By honest I don't mean that you will only tell what's true. But you make clear the entire situation. You make clear all the information that is required for somebody else who is intelligent to make up their mind.

– "The Unscientific Age," John Danz Lecture Series, 1963

Who are the witch doctors? Psychoanalysts and psychiatrists, of course. If you look at all of the complicated ideas that they have developed in an infinitesimal amount of time, if you compare to any other of the sciences how long it takes to get one idea after the other.

If you consider all the structures and inventions and complicated things, the ids and the egos, the tensions and the forces, and the pushes and the pulls, I tell you they can't all be there. It's too much for one brain or a few brains to have cooked up in such a short time.

– "The Unscientific Age," John Danz Lecture Series, 1963

Newspaper articles and popularization books tend to try to "explain" the latest views in a way that is simplified to be easy to understand, but what is understood then is wrong; sometimes only a little bit wrong but enough to throw you off.

– Letter to student Charles E. Tucker, April 1967

To find the proper place of scientific culture in modern society is not to solve the problems of modern society. There are a large number of problems that have nothing much to do with the position of science in society, and it is a dream to think, if you do think that at all, that to simply decide that one aspect, as to how ideally science and society should be matched, is somehow or other to solve all the problems.

– Galileo Symposium, "What Is and What Should Be the Role of Scientific Culture in Modern Society," September 1964

This world view all of us try from time to time to communicate to our unscientific friends and we get into difficulty most often because we get confused in trying to explain to them the latest questions, such as the meaning of the conservation of CP, whereas they don't know anything about the most preliminary things. For four hundred years since Galileo we have been gathering information about the world which they don't know.

– Galileo Symposium, "What Is and What Should Be the Role of Scientific Culture in Modern Society," September 1964

The great majority of people, the enormous majority of people, are woefully, pitifully, absolutely ignorant of the science of the world they live in, and they can stay that way, I don't mean that, I say to heck with them, what I mean is that they are able to stay that way without it worrying them at all — only mildly — so from time to time when they see CP* mentioned in the newspaper they ask what it is. An interesting question of the relation of science to modern society is just that — why is it possible for people to stay so woefully ignorant and yet reasonably happy in modern society when so much knowledge is missing to them?

– Galileo Symposium, "What Is and What Should Be the Role of Scientific Culture in Modern Society," September 1964

As I'd like to show Galileo our world, I must show him something with a great deal of shame. If we look away from the science and look at the world around us we find out something rather pitiful.

---

*CP stands for charge parity. Physicists once believed that the laws of physics for a particle and an antiparticle are the same as long as you make the particles symmetrical with respect to position, but this belief was found to be incorrect by James Cronin and Val Fitch in 1964, leading to the pair's winning the 1980 Nobel Prize in Physics.

That the environment that we live in is so actively, intensely unscientific.

– Galileo Symposium, "What Is and What Should Be the Role of Scientific Culture in Modern Society," September 1964

I believe that the science has remained irrelevant because we wait until somebody asks us questions or until we are invited to give a speech on Einstein's theory to people who don't understand Newtonian mechanics, but we never are invited to give an attack on faith-healing or on astrology on what is the scientific view of astrology today.

– Galileo Symposium, "What Is and What Should Be the Role of Scientific Culture in Modern Society," September 1964

I believe that we should demand that people try in their own minds to obtain for themselves a more consistent picture of their own world; that they not permit themselves in the luxury of having their brain cut in four pieces or two pieces even, and on one side they believe this and on the other side they believe that, but never try to compare the two points of view. Because we have learned that, by trying to put the points of view that we have in our head together and comparing one to the other, we make some progress in understanding and in appreciating where we are and what we are.

– Galileo Symposium, "What Is and What Should Be the Role of Scientific Culture in Modern Society," September 1964

I think we live in an unscientific age in which almost all the buffeting of communications and television words, books, and so on are unscientific. That doesn't mean they are bad, but unscientific. As a result, there is a considerable amount of intellectual tyrant in the name of science.

– National Science Teachers Association Fourteenth Convention lecture, "What Is Science?" April 1966

People think that experts know what they are doing. But most experts, whether in the stock market, education, sociology, or some parts of psychology, don't know more than the average person. They do studies, follow certain methods, and have results.

– *U.S. News and World Report* interview, February 1985

Publicity is an evil thing — once it starts to work — its practitioners don't know when to stop and it gets bigger and bigger.

– Letter to Jeanne Henry, July 1974

Although you say, "I'd love to get all the publicity we can stand," I personally cannot stand publicity and my wish would be the exact opposite. I'd love no publicity at all. We will both have to compromise a bit.

– Letter to Jeanne Henry, July 1974

There seems to be some interest in the world about how it is, how it feels and so on to do physics and so on. And various psychologists are always breathing down our neck to discover the mode, how creativity works.

– CERN talk, December 1965

I read your article very carefully, and I must say that the *Pakistan Observer* contained one of the best and clearest articles about what we did to win the Nobel Prize. All the big newspapers and news systems really do not carry anything like an attempt of the truth or explanation of the real situation. It takes a local paper and a man who really understands what things are about, to create a serious and sensible article that is not just full of pleasantries and trivia.

– Letter to Dr. A. M. Harun-ar Rashid, November 1965

It is unfair to use me to promote your own company.

– Letter to Jeanne Henry, July 1974

One of the ways of stopping science would be only to do experiments in the region where you know the law.

– *The Character of Physical Law*, p. 158

Right or wrong, I got the idea that the traditions of science are not — they're fragile. The traditions of scientific thought, I don't believe they're fragile. I think they're very fragile, and easily lost, and that science really has a value. The viewpoint that's involved, the objectivity, the way of doing things is valuable, see? So I thought it was of value, and that it might be destroyed. It could be destroyed because maybe people without this tradition would be the only ones left that would have any power.

– Interview with Charles Weiner, June 27, 1966 (Niels Bohr Library and Archives with the Center for the History of Physics)

People have this feeling that scientists know more about this than they do. It is not true. It's like the Wizard of Oz. You look behind and you see he's an ordinary fellow like you are.

– Interview with Charles Weiner, June 28, 1966 (Niels Bohr Library and Archives with the Center for the History of Physics)

If success resulted from using such a scientific attitude, a country might be encouraged to go on and develop a re-interest in the problems of science itself.

– MIT centennial, "Talk of Our Times," December 1961

I find that the lures that are put out in industry often exceed the reality.

– Interview with Charles Weiner, March 5, 1966 (Niels Bohr Library and Archives with the Center for the History of Physics)

Is science of any value? I think a power to do something is of value. Whether the result is a good thing or a bad thing depends on how it is used, but the power is a value.

– "The Uncertainty of Science," John Danz Lecture Series, 1963

Mathematics

I wish you ladies and gentleman out there knew some of this mathematics. It's not just the logic and accuracy of it all you're missing — it's the poetry too.

– BBC interview, "A Novel Force in Nature"

Take this neat little equation here. It tells me all the ways an electron can make itself comfortable in or around an atom. That's the logic of it. The poetry of it is that the equation tells me how shiny gold is, how come rocks are hard, what makes grass green, and why you can't see the wind. And a million other things besides, about the way nature works.

– BBC interview on the gauge theories, "A Novel Force in Nature"

It's not an accidental large number, like the large number which is the ratio of the volume of the earth to the volume of a flea.

– Audio recording of Feynman Lectures on Physics, Lecture 7, October 17, 1961

A vector is like a push that has a certain direction, or a speed that has a certain direction, or a movement that has a certain direction — and it's represented on a piece of paper by an arrow in the direction of the thing.

– *Feynman's Tips on Physics*, p. 23

You know, it's not true that what is called "abstruse" math is so difficult. Take something like computer programming, and the careful logic needed for that — the kind of thinking that mama and papa would have said was only for professors. Well, now it's part of a lot of daily activities, it's a way to make a living; their children get interested and get hold of a computer and they're doing the most crazy, wonderful things!

– *Omni* interview, February 1979

But you see, from the beginning I was disconnected. I was trying to find a formula for adding the integers together because I wanted the formula. I didn't care; it didn't mean anything to me, that this was worked out by the Greeks or even by the Babylonians in 2000 BC. This didn't interest me at all. It was my problem and I had fun out of it, you see. It was always that way. I was always playing my own independent game.

– Interview with Charles Weiner, March 4, 1966 (Niels Bohr Library and Archives with the Center for the History of Physics)

To those who do not know mathematics it is difficult to get across a real feeling as to the beauty, the deepest beauty, of nature. C. P. Snow talked about two cultures. I really think that those two cultures separate people who have and people who have not had this experience of understanding mathematics well enough to appreciate nature once.

– *The Character of Physical Law*, p. 58

Errors in algebra, differentiation, and integration are only nonsense; they're things that just annoy the physics, and annoy your mind while you're trying to analyze something. You should be able to do calculations as quickly as possible, and with a minimum of errors. That requires nothing but rote practice — that's the only way to do it.

– *Feynman's Tips on Physics*, p. 19

It is remarkable, but at this time it is possible to say that there is no experimental discrepancy at all between the physics theories anywhere and the results of the experiment. That doesn't mean we can compute everything. The rules of the game by which we make the computation are underneath everything. That's the way nature works. It's simple.

– "QED: Photons — Corpuscles of Light," Sir Douglas Robb Lectures, University of Auckland, June 1979

Now the idea is this: We want to see the momentum distribution of the partons inside. Imagine a swarm of bees coming toward you. Suppose you scatter radar from the swarm. The different bees would have different velocities, and when you scattered radar from them, you'd see a distribution of frequencies scattered from a monochromatic radio wave by a moving bunch of bees, you can get the distribution momentum of the bees inside the swarm.

– Caltech lecture on particles, 1973

Mathematics is a language plus reasoning; it is like a language plus logic. Mathematics is a tool for reasoning.

– *The Character of Physical Law*, p. 40

Euclid said, "There is no royal road to geometry." And there is no royal road. Physicists cannot make a conversation to any other language. If you want to learn about nature, to appreciate nature, it is necessary to understand the language that she speaks in. She offers her information in only one form; we are not so unhumble as to demand that she change before we pay attention.

– *The Character of Physical Law*, p. 58

So I got more and more interested in the mathematical business associated with physics. In addition, mathematics itself had a great appeal for me. I loved it all my life.

– Furture for Science interview

To use mathematics successfully one must have a certain attitude of mind — to know that there are many ways to look at any problem and at any subject.

– "New Mathematics," written for the California State Department of Education, 1965

You need an answer for a certain problem: the question is how to get it. The successful user of mathematics is practically an inventor of new ways of obtaining answers in given situations.

– "New Mathematics," written for the California State Department of Education, 1965

What is the best method to obtain the solution to a problem? The answer is, any way that works.

– "New Mathematics," written for the California State Department of Education, 1965

Mathematics is not a science from my point of view, in the sense that it's not a natural science. It's an unnatural science, perhaps.

– Audio recording of Feynman Lectures on Physics, Lecture 3, October 3, 1961

The final result of taking the limits is written still another way, as $\frac{ds}{dt}$. Now, it gets even more abstract. These ds are again not to be divided out. And if I tell you to call them an eensy weensy bit of time and distance, the mathematicians will get very angry at me. But it works fine. You can use it. If you imagine that these two are smaller than anything at all, then you've got the right idea.

– Audio recording of Feynman Lectures on Physics, Lecture 8, October 20, 1961

Now you're doing calculus. That's high-class calculus. Differentiating things. Nothing to it.

– Audio recording of Feynman Lectures on Physics, Lecture 8, October 20, 1961

You can work out your own table of integrals by differentiating until you're green around the gills, and you'll find for every formula with a differential, you've got an integral formula if you turn it around.

– Audio recording of Feynman Lectures on Physics, Lecture 8, October 20, 1961

Unlike the case of mathematics in which everything can be defined and then you don't know what you're talking about — in fact, the glory of mathematics is that you don't know what you're talking about. The glory is that the laws, the arguments, and the logic are independent of what it is.

– Audio recording of Feynman Lectures on Physics, Lecture 12, November 7, 1961

I never get my signs right — you may not either, but at the end you can figure it out.

– Audio recording of Feynman Lectures on Physics, Lecture 13, November 10, 1961

I'm always checking things while I calculate because I make so many mistakes. One way to check it is to do the mathematics very carefully; the other way to check it is to keep seeing whether the numbers that come out are sensible, whether they describe what's really happening.

– *Feynman's Tips on Physics*, p. 63

The apparent ease with which I do this is false: I swear I did it more than once before I got it right!

– *Feynman's Tips on Physics*, p. 62

Remember, whenever you're stuck in a mathematical analysis, you can always do it by arithmetic!

– *Feynman's Tips on Physics*, p. 82

I knew that one of the great problems of geometry was trisecting an angle. So the heck with learning all these little bits and pieces, we would go ahead and do the big problem, you see! That's a way to learn, though — I'm telling you.

– Interview with Charles Weiner, March 4, 1966 (Niels Bohr
Library and Archives with the Center for the History of Physics)

It wasn't anything brilliant, it's just that I was facile at mathematics, mathematical manipulating like a guy that's good at doing arithmetic fast in his head. That kind of stuff. It's very useful to be able to do.

– Interview with Charles Weiner, March 5, 1966 (Niels Bohr
Library and Archives with the Center for the History of Physics)

The only thing I could think of was, this stuff was good to teach somebody else, but it wasn't good itself. I still have that feeling about mathematics.

– Interview with Charles Weiner, March 5, 1966 (Niels Bohr
Library and Archives with the Center for the History of Physics)

I can't gamble. I understand the mathematics of the odds. I believe firmly that the games are presumably fair. They're honest. If they're honest, there's no game to it, because it's just a question of how the dice go, and it isn't interesting to me. It's just accident.

– Interview with Charles Weiner, June 27, 1966 (Niels Bohr
Library and Archives with the Center for the History of Physics)

I don't know why number theory does not find application in physics. We seem to need instead the mathematics of functions of continuous variables, complex numbers, and abstract algebra.

– Letter to Mr. Robert Boeninger, May 1969

I cannot understand mathematics very well as I have told you, and must have physical examples, but that is just the way my mind works.

– Letter to Bert and Mulaika Corben, 1948

Thinking I understand geometry and wanting to cut a piece of wood to fit the diagonal of a five-foot square, I try to figure out how long it must be. Not being very expert, I get infinity — useless nor does it help to say it may be zero because they are both circles. It is not philosophy we are after, but the behavior of real things. So in despair, I measure it directly — lo, it is near to seven feet — neither infinity nor zero. So, we have measured these things for which our theory gives us numbers close to what we measure. We are seeking the formula that gives the square root of fifty.

– Letter to Barbara Kyle, October 1965 (*Perfectly Reasonable Deviations from the Beaten Track*, p. 152)

I am not interested in what today's mathematicians find interesting.

– Letter to Dr. John A. Wheeler, May 1966

Set theory is used sometimes, but not very often, in practical matters. Its greatest uses and beauty appears in the study of the logical foundations of mathematics.

– Letter to Alexander Calandra, September 1965

A pure mathematician is very impractical — he is not interested (in fact, he is purposely disinterested) in the meaning of the mathematical symbols and letters and ideas, but is only interested in logical interconnection of the axioms, while the user of mathematics has to understand the connection of mathematics to the real world.

– "New Mathematics," written for the California State Department of Education, 1965

The successful user of mathematics is practically an inventor of new ways of obtaining answers in given situations.

– "New Mathematics," written for the California State Department of Education, 1965

There are many ways of doing a problem but, unfortunately, there also are definite known fixed ways of doing the problems. What we have been doing in the past is teaching just one fixed way to do arithmetic problems instead of teaching the flexibility of mind; that is, the various possible ways of writing down a problem, the possible ways of thinking about it, and the possible ways of getting at the problem.

– "New Mathematics," written for the California State Department of Education, 1965

In order to use mathematics one must have a deeper understanding of the relation of mathematical objects to real things and this deeper understanding is opposed to a tendency to try to make all of the different kinds of mathematical applications represent the same thing.

– "New Mathematics," written for the California State Department of Education, 1965

Strictly and technically, if it be known, the theory of the fundaments of mathematics is not in a completely satisfactory state and involves some very great complications.

  – "New Mathematics," written for the California State Department of Education, 1965

I might, perhaps, mention why some attention is paid to different base systems besides base 10. The point is only to increase again mathematical experience. To make clear to children that the base 10 is an accident of human history and that there is nothing special about the number 10.

  – "New Mathematics," written for the California State Department of Education, 1965

If I were giving a talk on mathematics, I would have already answered you. Mathematics is looking for patterns.

  – National Science Teachers Association Fourteenth Convention lecture, "What Is Science?" April 1966

So he knew all his arithmetic, and he was very good at it, and that was a challenge to me. I kept practicing. We used to have a little contest. Every time we'd have to calculate anything, we'd race to the answer, he and I, and I would lose. After several years I began to get in there once in a while, maybe one out of four. You have to notice the numbers, you see — and each of us would notice a different way. We had lots of fun.

  – On his experience with the Manhattan Project, "Los Alamos from Below," 1975

Of course you'd notice something funny about a number like if you have to multiple 174 by 140, for example. You notice that 173 by 141 is like the square root of 3, times the square root of 2, which is the square root of 6, which is 245. But you have to notice the numbers, you see, and each guy would notice a different way — we had lots of fun.

– UCSB talk, "Los Alamos from Below," February 1975

There were these high priests that could predict Venus, and the normal men, who you will imagine said, said, "God, how do they do this? This is marvelous, it frightens me, I can't learn arithmetic, this is impossible," and so forth. Now if there had been in those days an attempt at pedagogy, in other words, the priests trying to explain to the laymen what they were doing, they could explain they were only counting or doing the equivalent of counting.

– Esalen lecture, "Quantum Mechanical View of Reality (Part 2)," October 1984

I will show you how to make these calculations as if you were taking beans out of pots!

– Esalen lecture, "Quantum Mechanical View of Reality (Part 2)," October 1984

I think we've understood all we can from that point of view; what we've found in this century is different enough, obscure enough, that further progress will require a lot of math.

– *Omni* interview, February 1979

I don't believe in the idea that there are a few peculiar people capable of understanding math, and the rest of the world is normal. Math is a human discovery, and it's no more complicated than humans can understand.

  – *Omni* interview, February 1979

Math is just a language.

  – In personal notes

Math may permit wildly different apparent starting points.

  – In personal notes

The rules of algebra are things studied by mathematicians, and mathematicians have tried to find all the objects you could possibly find which obey those rules! The rules were originally made for counting apples; it was improved by using negative numbers; it was improved still further by inventing fractions.

  – "QED: Fits of Reflection and Transmission," Sir Douglas Robb
    Lectures, University of Auckland, 1979

[On complex numbers:] It's interesting that mathematicians all this mathematics for these crazy numbers without having anything to directly apply it to in physics and that it should turn out to be so fundamental to the bottom laws of physics using such funny numbers.

  – "QED: Fits of Reflection and Transmission," Sir Douglas Robb
    Lectures, University of Auckland, 1979

We have, today, become very sophisticated. In the early days, when mathematics was first developing and it was said that a number is something like when you count the number of apples or people or something like that, then the idea of half of a person was somewhat of a problem. But nowadays there's no difficulty at all, and nobody has any moral or discomforting, gory feelings when they hear there are 3.2 people per square mile in a certain region.

– "QED: Fits of Reflection and Transmission," Sir Douglas Robb Lectures, University of Auckland, 1979

You'd be surprised how many numbers you could make by playing with pi's and 2s and 5s and so on, and if you haven't got anything to guide you except the answer, you can make it come out to even several decimal places by suitable jiggling about. It's surprising how close you can make an arbitrary number by playing around with nice numbers like pi.

– "QED: New Queries," Sir Douglas Robb Lectures, University of Auckland, 1979

Now, you say, is that the correct way to do it? There is no such thing! There is no "correct" way to do anything. A particular way of doing it may be correct, but it is not the correct way.

– *Feynman's Tips on Physics*, p. 58

On looking back over the work, I can only feel a kind of regret for the enormous amount of physical reasoning and mathematically re-expression which ends by merely re-expressing what was previously known, although in a form which is much more efficient for the calculation of specific problems. Would it not have been much easier to simply work entirely in the mathematical framework to elaborate a more efficient expression? This would certainly seem to be the case, but it must be remarked that although the problem actually solved was only such a reformulation, the problem originally tackled was the (possibly still unsolved) problem of avoidance of the infinities of the usual theory. Therefore, a new theory was sought, not just a modification of the old. Although the quest was unsuccessful, we should look at the question of the value of physical ideas in developing a new theory.

– From *Nobel Lectures, Physics 1963–1970*, Elsevier Publishing Company, Amsterdam, 1972

When I was in high school, we had an algebra team, and I was on the algebra team. It was a crazy thing, where we'd meet together with another school, and they'd open from an envelope problems that somebody invented somewhere, and they'd announce, "This problem is 45 seconds," or maybe "two and a half minutes," and they would write it on the board. You had 15 seconds to think and then you'd work like a demon, and you'd put a circle around your answer. It didn't make any difference how you got the answer. This is what I loved. I used to practice, to do this, and I would get very, very fast. The ability to do algebra fast, which later became the ability to do calculus fast, always stood me in good stead.

– Interview with Charles Weiner, March 5, 1966 (Niels Bohr Library and Archives with the Center for the History of Physics)

It seems to me if you say $2x = 32$, I know what it means. x always represents some number, so if the number happens to be up there as an exponent, does it make any difference? I mean, we knew about exponents. Does it make any difference the problem?

– Interview with Charles Weiner, March 4, 1966 (Niels Bohr Library and Archives with the Center for the History of Physics)

When I was at MIT, I would read books on fields that I didn't know, like general relativity and so on — or even in the encyclopedia, when I'd pull stuff out of the article — I seemed to have a sense to pull, in electrostatics, a lot of stuff, but when it went to the calculation of the capacitance of an elliptical condenser, which was quite complicated, it didn't bother me that I didn't understand it. I knew that was not so interesting as the general theorems about the laws of the inverse square, and, you know, I had some way of knowing what was important and what was not.

– Interview with Charles Weiner, March 4, 1966 (Niels Bohr Library and Archives with the Center for the History of Physics)

My father taught me about pi before I had learned in the school the decimals for fractions, and explained decimals. See, I was really ahead in arithmetic. I remember him telling me about pi as a great and marvelous mystery. Everything was always dramatic — that all circles have the same ratio of the distance around to the distance across, and that this number, this strange number, is of very great significance, and is a marvelous number. So pi was like in gold letters, you see.

– Interview with Charles Weiner, March 4, 1966 (Niels Bohr Library and Archives with the Center for the History of Physics)

Pure mathematics is just such an abstraction from the real world, and pure mathematics does have a special precise language for dealing with its own special and technical subjects. But this precise language is not precise in any sense if you deal with real objects of the world, and it is only pedantic and quite confusing to use it unless there are some special subtleties which have to be carefully distinguished.

– "New Mathematics," written for the California State Department of Education

Technology

There is a computer disease that anybody who works with computers knows about. It's a very serious disease and it interferes completely with the work. The trouble with computers is that you "play" with them!

— *Surely You're Joking, Mr. Feynman!*, p. 127

The right answer must be determined by experiment, and pure speculation has no place — but not being able to experiment I could not keep my mind from wandering on the plane. So I should like to report on my pure speculations just for fun, not in the sense that I think these things are necessarily so, but only for the fun of thinking how exciting it might be.

> – Letter Edwin H. Land (Polaroid Corporation), May 1966
> (*Perfectly Reasonable Deviations from the Beaten Track*, p. 221)

Computers, however, can't yet "catch on" to what is being said, the way a person does. They need to be told in excruciating detail exactly what to do. Perhaps one day we will have machines that can cope with approximate task descriptions, but in the meantime we have to be very prissy about how we tell computers to do things.

> – *Feynman Lectures on Computation*, p. 3

[On quantum computers:] What we really need to worry about is not the absolute energy, but the energy that's lost — that is, the free energy that produces chaos or irregularities. This concern is just like what somebody said to me about my atomic computer. You've estimated the energy, he said, but you've forgotten how much energy it takes to manufacture the computer. Well fine. Let's say that's the mc^2 of all the atoms. But when I'm done computing after 100 years, I'm going to get that energy back again.

> – "Tiny Computers Obeying Quantum Mechanical Laws,"
> *New Directions in Physics: The Los Alamos 40th Anniversary Volume*, 1987

I want to talk about the possibility that there is to be an exact simulation, that the computer will do exactly the same as nature. If this is to be proved and the type of computer is already explained, then it's going to be necessary that everything that happens in a finite volume of space and time would have to be exactly analyzable with a finite number of logical operations.

– "Simulating Physics with Computers," *International Journal of Theoretical Physics*, May 1981

The discovery of computers and the thinking of computers have turned out to be extremely useful in many branches of human reasoning. For instance, we never really understood how lousy our understanding of language was, the theory of grammar, and all that stuff, until we tried to make a computer work which would be able to understand language.

– MIT conference, May 1981

[On the *Challenger* explosion:] The fact that this danger did not lead to catastrophe before is no guarantee it will not the next time, unless it is completely understood.

– "Feynman Takes NASA to Task," *Pasadena Star-News*, June 11, 1986

A computer is a high-class, super-speed, nice, streamlined filing clerk.

– Esalen lecture, "Computers from the Inside Out," October 1984

It is always better to make the machines smaller, and the question is how much smaller is it still possible to make machines according to the laws of Nature, in principle. I will not discuss which and what of these things will actually appear in the future. That depends on economic problems and social problems, and I am not going to try to guess at those.

– "The Computing Machines in the Future," Nishina Memorial Lecture, August 1985

The idea of the telescope is to focus the light from a bigger area into a smaller area, so we can see things that are weaker, less light.

– BBC, "Fun to Imagine" television series, 1983

Good machining is essential to building good apparatus for the precise and careful measurements required in physics to discover Nature's Laws.

– Letter to Raymond Rogers, January 1966 (*Perfectly Reasonable Deviations from the Beaten Track*, p. 208)

There is a great deal of work to try to develop smarter machines, machines which have a better relationship with the humans so that input and output can be made with less effort than the complex programming that's necessary today. This goes under the name of artificial intelligence, but I don't like that name.

Perhaps the unintelligent machines can do even better than the intelligent ones.

– "The Computing Machines in the Future," Nishina Memorial Lecture, August 1985

I mentioned something about the possibility of time — of things being affected not just by the past but also by the future, and therefore that our probabilities are in some sense "illusionary." We have only the information from the past, and we try to predict the next step, but in reality it depends on the near future, which we can't get at, or something like that.

– MIT conference, May 1981

For fun again and intellectual pleasure, we could imagine machines tiny like few microns across wheels and cables all interconnected by wires, silicon connections, so that the thing as a whole, a very large device, moves not like the awkward motions of our present stiff machines but in a smooth way of the neck of a swan, which after all is a lot of little machines, the cells all interconnected and all controlled in a smooth way. Why can't we do that ourselves?

– "The Computing Machines in the Future," Nishina Memorial Lecture, August 1985

[On investigating the *Challenger* explosion:] The main thing I learned at that meeting was how inefficient a public inquiry is: Most of the time, other people are asking questions you already know the answer to — or are not interested in — and you get so fogged out that you're hardly listening when important points are being passed over.

> – "Feynman: Frustrated by the Slow Pace of Probe," *Pasadena Star-News*, January 29, 1989

It is not necessary to understand the way birds flap their wings and how the feathers are designed in order to make a flying machine. It is not necessary to understand the lever system in the legs of a cheetah, that is, an animal that runs fast, in order to make an automobile with wheels that go very fast. It is, therefore, not necessary to imitate the behavior of Nature in detail in order to engineer a device which can in many respects surpass Nature's abilities.

> – "The Computing Machines in the Future," Nishina Memorial Lecture, August 1985

Have you ever asked yourself what you are made of? If you answer, flesh and bones, then I'll ask what they're made of — you'll say molecules, proteins, perhaps even DNA. What are those molecules made of — atoms: We are all a pile of atoms.

> – BBC, "Horizon: The Hunting of the Quark," May 1974

War

There is a responsibility for all of us for the way the world has come out, but we didn't have enough sense to think of a way of avoiding it.

– "900 at Caltech, JPL Declare Support for Nuclear Arms Freeze," *The Los Angeles Times*, October 16, 1982

People said, "Don't you feel guilty?" afterwards. No, because at that moment is when I did my thinking of the seriousness of the situation in the world, of the possibilities of making a bomb, of the dangers that would result if the other side were to have done it if we didn't, and it seemed to me absolutely clear that we must really make an effort to save the world from the other possibility. It's true, it turned out later that they weren't doing very well, but there was no way at the time to know, because it was possible. We showed it was possible. It was not impossible, therefore, that they would do it, and if they did, that would be terrible.

– Interview with Charles Weiner, March 5, 1966 (Niels Bohr Library and Archives with the Center for the History of Physics)

On the other hand, as a possibility of a large release of power, now it's often said that science makes the bomb. It isn't. It's engineering which makes the bomb. The reason to make the bomb was a military reason during the war. It's true that scientists worked on it, but they weren't working on science, they were being engineers during the war. They were taken away from the laboratories in order to do this.

– Interview for Viewpoint

A further result of the calamity of war might be a universal antagonism toward physics, as a result of the destruction which people might blame the scientists for having made possible.

– MIT centennial, "Talk of Our Times," December 1961

I would like an opportunity to describe, instead of the horror of the things I helped to make, rather some of the hopes and values also created which might arise if the war problem could be solved.

– Notes on the atom bomb

The practical attainment during the war of the transmutation of the elements and the release of subatomic energy has once again emphasized that the discoveries of science are never good or bad in themselves. It is a question of what use we make of the discoveries. Science gives us power. We can utilize it either to further good or to further evil. And the release of nuclear energy is the release of a most tremendous power. It is potentially of extreme value, or as we have seen, is capable of unprecedented destruction.

– Notes from Los Alamos

A discussion of the future of atomic energy is a discussion of the problems arising in making the choices.

– Notes from Los Alamos

Because knowledge can be applied to war and war is bad is no reason to seek to stop war by suppressing knowledge. Even if we kept science going here and not leaking out, we couldn't prevent its advance in other hands. We sit, others act.

– Notes from Los Alamos

Most bitter debate centers around control of information, development of military weapons and research into new knowledge.

– Notes from Los Alamos

Problem is science gives power for good or evil. We are concerned it may be used for evil. Monopolization by the military does not strike any sane man that the choice is to develop the good ends — but is forcing science to develop in just those directions in which it can be made most destructive.

– Notes from Los Alamos

Scientists discovered energy locked in atoms and dreamed of the age of its release. The great atomic age was ushered in in a mathematic and horrible fashion. The greatest problem by all odds is that of war and peace.

– Notes

I don't know myself whether I am for nuclear testing or against nuclear testing. There are reasons on both sides. It makes radioactivity, and it's dangerous, and it's also very bad to have a war. But whether it's going to be more likely to have a war because you test, I don't know. Whether preparation will stop the war, or lack of preparation, I don't know.

– "The Unscientific Age," John Danz Lecture Series, 1963

All the equipment from different research was being put together to make the new apparatus to do the experiment to try to separate the isotopes of uranium.

– UCSB talk, "Los Alamos from Below," February 1975

Bomb has great danger — war — any nation can and will develop atomic power.

– Notes from before Los Alamos

It is the lesson of the last war not to think of people as having special inherited attributes simply because they are born from particular parents, but to try to teach these "valuable" elements to all men because all men can learn, no matter what their race.

– Letter to Tina Levitan, February 1967 (*Perfectly Reasonable Deviations from the Beaten Track*, p. 235)

Although in my field at the present time I'm a slightly famous man, at the time I was not anybody famous at all. I did not even have a degree when I started to work on my stuff associated with the Manhattan Project.

– UCSB talk, "Los Alamos from Below," February 1975

All the science stopped during the war except the little bit that was done in Los Alamos.

– UCSB talk, "Los Alamos from Below," February 1975

All the things they'd thought of in Berkeley about the atomic bomb, and nuclear physics and all these things, and I didn't know very much about it. I had been doing other kinds of things. And so I had to do an awful lot of work.

– UCSB talk, "Los Alamos from Below," February 1975

I never traveled on an airplane before; I traveled on an airplane. They strapped the secrets, with a little thing with a belt on my back!

– UCSB talk, "Los Alamos from Below," February 1975

I have a great deal of respect for these military guys because I never can decide anything very important in any length of time at all.

– UCSB talk, "Los Alamos from Below," February 1975

We always were in a hurry. I have to explain that everything we did, we tried to do as quickly as possible.

– UCSB talk, "Los Alamos from Below," February 1975

I told them what we were doing, and they were all excited. We're fighting a war. We see what it is. They knew what the numbers meant. If the pressure came out higher that meant there was more energy released and so on and so on. They knew what they were doing.

– UCSB talk, "Los Alamos from Below," February 1975

I'd sit in a restaurant in New York, for example, and looked at the buildings and how far away, I would think, you know, how much the radius of the Hiroshima bomb damage was and so forth. How far down there was down to 34th Street. All those buildings, all smashed and so on. And I got a very strange feeling. I would go along and I would see people building a bridge. Or they'd be making a new road, and I thought, they're crazy, they just don't understand, they don't understand. Why are they making new things, it's so useless?

– UCSB talk, "Los Alamos from Below," February 1975

The attitude is that this business about how you open safes is not something that everybody should know because it makes everything very unsafe, it's very dangerous to have everybody know how to do this.

– UCSB talk, "Los Alamos from Below," February 1975

I worked on the Los Alamos project during World War II. After the first successful test there was tremendous excitement. Everyone had parties; we all ran around. I sat on the end of a jeep and beat drums. One man, Bob Wilson, was just sitting there and moping. When I asked him why, he said, "It's a terrible thing that we made."

– *U.S. News and World Report* interview, February 1985

[On the atomic bomb:] Maybe, by some miracle, those who have the responsibility to control this thing will begin to realize — or maybe already realize — its uselessness. In that case, the bomb may turn out to have been useful in stopping us from the age-old historical exercise of destroying each other.

– *U.S. News and World Report* interview, February 1985

[On the Los Alamos project:] Suppose scientists hadn't gone ahead and, instead, had said, "It will be such a serious problem for humankind later that we don't think that we should do it." What a scream there would have been if Hitler and his crew had managed to make the bomb and used it to dominate the world.

– *U.S. News and World Report* interview, February 1985

It's very annoying to have everybody ask you for an opinion, and you're supposed to be wise all of a sudden, and I know I'm not wise all of a sudden.

– "The Remarkable Dr. Feynman," *Los Angeles Times Magazine*,
April 20, 1986, p. 37

I made the decision to work on the Manhattan Project at the beginning of the war because I thought the Germans would do it. I don't know whether that was the right decision.

– "The Remarkable Dr. Feynman," *Los Angeles Times Magazine*,
April 20, 1986

From a scientific point of view, the Manhattan Project was not what I would have ordinarily wanted to do; it was engineering more than science. It was very exciting to meet all the great men and smart characters that you read about. It's a similar response I have to the commission. I wouldn't have wanted to do the commission aside from the feeling of duty, but once I decide that I'm stuck and I've got to do it, then I've got to work hard. But if you gave me half a chance, I'd quit. It's exciting, once you're stuck in it. It's like asking somebody who is almost having an automobile accident whether it's exciting. You're damn tootin' it's fun trying to steer between the cars, isn't it? Only he'd rather not have to do that.

– "The Remarkable Dr. Feynman," *Los Angeles Times Magazine*,
   April 20, 1986, p. 37

[Opposing bilateral nuclear arms freeze:] [Reagan is one of the] experts, especially in government, who don't know what they're doing.

– Quoted in *Star-News*, October 16, 1982

The terror and danger of the atomic bomb has been told over many times.

– Notes

The atomic bomb is essentially just a very big bomb 1000 times more energetic than the biggest block buster.

– Notes

A nation will very soon be able to declare the death sentence to half the population of any other nation and within the day carry out the execution.

– Notes

Slight hope that it might be blessing in horrible disguise by bringing the lukewarm idealism toward cooperation between people into the reality born of necessity

– Notes

It has always been clear that cooperation was very desirable.

– Notes

The slight hope that the horror of the atomic bomb could possibly finally convince mankind the folly of division.

– Notes

[On the Los Alamos project:] What happened to me — what happened to the rest of us — is we started for a good reason, then you're working very hard to accomplish something, and it's pleasure, it's excitement. And you stop thinking, you know; you just stop.

– "The Feynman Legend," *The Los Angeles Times*, February 17, 1988

This same business with the bomb and this pessimism kept with me for several years and, by 1950, I still was pessimistic about the world and was pretty sure that I had it right, that nobody was getting anywhere and we were all going around in circles and that we were going to have trouble. Then when we'd have trouble with Russia and so on, we'd bomb each other out, and the Northern Hemisphere would be in a bad way.

– Interview with Charles Weiner, June 27, 1966 (Niels Bohr Library and Archives with the Center for the History of Physics)

[On the nuclear bomb:] As far as I know — maybe other people don't agree — there was in my experience no serious difficulty produced because information was maintained as secret that was essential to a more or less fundamental understanding, or was kept secret too long. There were important things which were released gradually — but in time, so to speak.

– Interview with Charles Weiner, June 27, 1966 (Niels Bohr Library and Archives with the Center for the History of Physics)

We scientists are clever — too clever — are you not satisfied? Is four square miles in one bomb not enough? Men are still thinking. Just tell us how big you want it!

– Notes

And that's one thing I did learn, that if you have some reason for doing something that's very strong and you start working on it, you must look around every once in a while and find out if the original motives are still right.

– Future for Science interview

Challenger

RICHARD FEYNMAN

Nothing short of a subpoena from Congress will get me to Washington again.

> – Letter to David Acheson, 1986 (*Perfectly Reasonable Deviations from the Beaten Track*, p. 405)

When using a mathematical model, careful attention must be given to uncertainties in the model.

– *Star-News*, Opinion, June 18, 1986

[On the Rogers Commission:] I feel like a bull in a china shop. The best thing is to put the bull out to work the plow. A better metaphor will be an ox in a china shop, because the china is the bull, of course.

– Letter to Gweneth and Michelle Feynman, February 12, 1986 (*Perfectly Reasonable Deviations from the Beaten Track*, p. 402)

If in this way the government would not support them, then so be it.

– Report of the Presidential Commission on the Space Shuttle *Challenger* Accident, Volume 2: Appendix F, June 1986

NASA owes it to the citizens from whom it asks support to be frank, honest, and informative, so that these citizens can make the wisest decisions for the use of their limited resources.

– Report of the Presidential Commission on the Space Shuttle *Challenger* Accident, Volume 2: Appendix F, June 1986

[On the Rogers Commission:] I'm a little disturbed that I felt a sense of duty.

 – "The Remarkable Dr. Feynman," *Los Angeles Times Magazine*,
    April 20, 1986

[On the Rogers Commission:] During the discussion earlier there are various pious remarks about how we as individuals, or better, small groups (called subcommittees), can go anywhere we want to get info. I try to propose I do that (and several physicists tell me they would like to go with me) and I have set my affairs so I can work intensively full time for a while. I can't seem to get an assignment, and the meeting breaks up practically while I am talking, with the Vice Chairman (Armstrong) remark about our not doing detailed work.

 – Letter to Gweneth and Michelle Feynman, February 12, 1986
    (*Perfectly Reasonable Deviations from the Beaten Track*, p. 399)

Sunday, I go with Graham and his family to see the Air and Space museum which Carl liked so much — we are in an hour before the official opening and there are no crowds — influence; after all, the acting head of NASA.

 – Letter to Gweneth and Michelle Feynman, February 12, 1986
    (*Perfectly Reasonable Deviations from the Beaten Track*, p. 401)

[On Rogers's leniency with fraternizing with the press:] I was pleased by his reaction, but now as I write this, I have second thoughts. It was too easy after he explicitly talked about the importance of no leaks, etc. at earlier meetings. Am I being se up? (See, darling, Washington Paranoia is setting in.) If, when he wants to stop or discredit me, he could charge me with leaking something important. I think it possible that there are things in this that somebody might be trying to keep me from finding out and might try to discredit me if I get too close.

– Letter to Gweneth and Michelle Feynman, February 12, 1986
  (*Perfectly Reasonable Deviations from the Beaten Track*, p. 401)

[On Rogers:] I am probably a thorn in his side.

– Letter to Gweneth and Michelle Feynman, February 12, 1986
  (*Perfectly Reasonable Deviations from the Beaten Track*, p. 401)

[On the Rogers Commission:] At this rate we will never get down close enough to business to find out what happened.

– Letter to Gweneth and Michelle Feynman, February 12, 1986
  (*Perfectly Reasonable Deviations from the Beaten Track*, p. 401)

[On the Rogers Commission:] I am determined to do the job of finding out what happened — let the chips fall.

– Letter to Gweneth and Michelle Feynman, February 12, 1986
  (*Perfectly Reasonable Deviations from the Beaten Track*, p. 401)

[On the Rogers Commission:] My guess is that I will be allowed to do this, overwhelmed with data and details, with the hope that so buried with all attention on technical details I can be occupied, so they have time to soften especially dangerous witnesses, etc. But it won't work because (1) I do technical information exchange and understanding much faster than they imagine, and (2) I already smell certain rats that I will not forget because I just love the smell of rats.

– Letter to Gweneth and Michelle Feynman, February 12, 1986
(*Perfectly Reasonable Deviations from the Beaten Track*, p. 402)

Official management, on the other hand, claims to believe the probability of failure is a thousand times less. One reason for this may be an attempt to assure the government of NASA perfection and success in order to ensure the supply of funds. The other may be that they sincerely believed it to be true, demonstrating an almost incredible lack of communication between themselves and their working engineers.

– Report of the Presidential Commission on the Space Shuttle
Challenger Accident, Volume 2: Appendix F, June 1986

It appears that there are enormous differences of opinion as to the probability of a failure with loss of vehicle and of human life. The estimates range from roughly 1 in 100 to 1 in 100,000. The higher figures come from the working engineers, and the very low figures from management. What are the causes and consequences of this lack of agreement? Since 1 part in 100,000 would imply that one could put a Sshuttle up each day for 300 years expecting to lose only one, we could properly ask "What is the cause of management's fantastic faith in the machinery?"

– Report of the Presidential Commission on the Space Shuttle
Challenger Accident, Volume 2: Appendix F, June 1986

The argument that the same risk was flown before without failure is often accepted as an argument for the safety of accepting it again. Because of this, obvious weaknesses are accepted again and again, sometimes without a sufficiently serious attempt to remedy them, or to delay a flight because of their continued presence.

– Report of the Presidential Commission on the Space Shuttle *Challenger* Accident, Volume 2: Appendix F, June 1986

It would appear that, for whatever purpose, be it for internal or external consumption, the management of NASA exaggerates the reliability of its product, to the point of fantasy.

– Report of the Presidential Commission on the Space Shuttle *Challenger* Accident, Volume 2: Appendix F, June 1986

The O-rings of the Solid Rocket Boosters were not designed to erode. Erosion was a clue that something was wrong. Erosion was not something from which safety can be inferred.

– Report of the Presidential Commission on the Space Shuttle *Challenger* Accident, Volume 2: Appendix F, June 1986

An entirely automatic landing is probably not as safe as a pilot-controlled landing.

– Report of the Presidential Commission on the Space Shuttle *Challenger* Accident, Volume 2: Appendix F, June 1986

We find that the attitude to system failure and reliability is not nearly as good as for the computer system. For example, a difficulty was found with certain temperature sensors sometimes failing. Yet 18 months later, the same sensors were still being used, still sometimes failing, until a launch had to be scrubbed because two of them failed at the same time.

– Report of the Presidential Commission on the Space Shuttle
Challenger Accident, Volume 2: Appendix F, June 1986

The action of the jets is checked by sensors, and, if they fail to fire the computers choose another jet to fire. But they are not designed to fail, and the problem should be solved.

– Report of the Presidential Commission on the Space Shuttle
Challenger Accident, Volume 2: Appendix F, June 1986

If a reasonable launch schedule is to be maintained, engineering often cannot be done fast enough to keep up with the expectations of originally conservative certification criteria designed to guarantee a very safe vehicle. In these situations, subtly, and often with apparently logical arguments, the criteria are altered so that flights may still be certified in time. They therefore fly in a relatively unsafe condition, with a chance of failure of the order of a percent.

– Report of the Presidential Commission on the Space Shuttle
Challenger Accident, Volume 2: Appendix F, June 1986

Let us make recommendations to ensure that NASA officials deal in a world of reality in understanding technological weaknesses and imperfections well enough to be actively trying to eliminate them.

– Report of the Presidential Commission on the Space Shuttle
Challenger Accident, Volume 2: Appendix F, June 1986

We have also found that certification criteria used in Flight Readiness Reviews often develop a gradually decreasing strictness. The argument that the same risk was flown before without failure is often accepted as an argument for the safety of accepting it again. Because of this, obvious weaknesses are accepted again and again, sometimes without a sufficiently serious attempt to remedy them, or to delay a flight because of their continued presence.

– *Pasadena Star-News*, Opinion, June 18, 1986

Out of a total of nearly 2,900 flights, 121 failed (1 in 25). This includes, however, what may be called early errors, rockets flown for the first few times in which design errors are discovered and fixed. A more reasonable figure for the mature rockets might be 1 in 50. With special care in the selection of the parts and in inspection, a figure of below 1 in 100 might be achieved but 1 in 1,000 is probably not attainable with today's technology. NASA officials argue the figure is much lower.

– *Pasadena Star-News*, Opinion, June 18, 1986

In fact, previous NASA experiments had shown, on occasion, just such difficulties, near accidents and accidents, all giving warning that the probability of flight failure was not so very small.

– *Pasadena Star-News*, Opinion, June 18, 1986

Why do we find such an enormous disparity between the management estimate and the judgment of the engineers? It would appear that, for whatever purpose, be it for internal or external consumption, the management of NASA exaggerates the reliability of its product, to the point of fantasy.

– *Pasadena Star-News*, Opinion, June 18, 1986

The fact that this danger did not lead to a catastrophe before is no guarantee that it will not the next time, unless it is completely understood. When playing Russian roulette, the fact that the first shot went off safely is little comfort for the next.

– *Pasadena Star-News*, Opinion, June 18, 1986

I report to Rogers that I have these close relatives with press connections and is it OK to visit with them? He is very nice and says, "Of course."

– Letter to Gweneth and Michelle Feynman, February 12, 1986
 (*Perfectly Reasonable Deviations from the Beaten Track*, p. 401)

I am not interested in my personal comfort; rather the well being of the country.

– Correspondence with Rogers ("Mr. Feynman Goes to Washington," 1986)

Politics

We all know that they don't know what they're doing in Washington. It's not that they're fools; it's just that nobody knows how to handle many of the problems. A lot of experts have studied these subjects. But they know much less than they will admit. If somebody ran for office saying that they didn't have answers, nobody would pay any attention to him. Everybody wants an answer. But someday, maybe, everybody will slowly come around to the realization that the experts don't know almost everything.

– *U.S. News and World Report* interview, February 1985

If it can be controlled in thermonuclear reactions, then it turns out that the energy that can be obtained from a quart of water a second is equal to the all the electrical power generated in the United States. With a faucet running at 15 gallons of water a minute, you have the fuel to supply all the energy in the United States that's used today. Therefore, it's up to the physicist to figure out how to liberate us from the need for having energy. And in can be done, in practice.

– Audio recording of Feynman Lectures on Physics, Lecture 4, October 6, 1961

[On the electron's negativity:] It's the scale I want to measure, and I use the "minus" because of Benjamin Franklin, who chose to call the electron "minus." Okay? So we're stuck with that since 1776. And lots of other things we're stuck with since 1776. And some of them they're not so concerned as I am.

– "QED: New Queries," Sir Douglas Robb Lectures, University of Auckland, 1979

Science is an international human effort, and there would be no "American science" were it not for scientific development in the rest of the world.

– Letter to Mr. Stuart Zimmer, February 1982 (*Perfectly Reasonable Deviations from the Beaten Track*, p. 344)

This is in the attitude of the mind of the populace, that they have to have an answer and that a man who gives an answer is better than a man who gives no answer, when the real fact of the matter is, in most cases it is the other way around. And the result of this is that the politician must give an answer. And the result of this is that political promises can never be kept. It is a mechanical fact, it is impossible. The result of that is that nobody believes campaign promises. And the result of that is a general disparaging of politics, a general lack of respect for the people who are trying to solve problems.

– "The Unscientific Age," John Danz Lecture Series, 1963

They decided to do something utterly illegal, which was to censor mail of people inside the United States, in the Continental United States, which they have no right to do. So it had to be set up very delicately, as a voluntary thing. We would all volunteer, not to seal our envelopes that we would send our letters out with. We would accept, it would be all right, that they would open letters coming in to us; that was voluntarily accepted by us.

– UCSB talk, "Los Alamos from Below," February 1975

For a very quick summary, I see nothing wrong with nuclear power except questions of the possibility of explosions, sabotage, stealing fuel to make bombs, leaking stored radioactive spent rods, etc. But all these are technical or engineering questions, about which we can do a great deal.

– Letter to student Mark Minguillon, April 1976 (*Perfectly Reasonable Deviations from the Beaten Track*, pp. 304–305)

The television industry can be proud to be a part of the tradition of freedom of expression of this country.

– Letter to Bill Whitley (KNXT), May 1959 (*Perfectly Reasonable Deviations from the Beaten Track*, p. 101)

The government of the United States is not very good, but it, with the possible exception of the government of England, is the greatest government on the earth today, is the most satisfactory, the most modern, but not very good.

– "The Uncertainty of Values," John Danz Lecture Series, 1963

I don't think of the problem as between socialism and capitalism but rather between suppression of ideas and free ideas. If it is free ideas and socialism is better than communism, it will work its way through. And it will be better for everybody. And if capitalism is better than socialism, it will work its way through.

– "The Uncertainty of Values," John Danz Lecture Series, 1963

According to the Constitution there are supposed to be votes. It isn't supposed to be automatically determinable ahead of time on each of the items what's right and what's wrong. Otherwise there wouldn't be the bother to invent the Senate to have the votes. As long as you have the votes at all, then the purpose of the votes is to try to make up your mind which is the way to go.

– "The Unscientific Age," John Danz Lecture Series, 1963

There is a system of laws and juries and judges. And although there are of course many faults and flaws, and we must continue to work on them, I have great admiration for that.

– "The Unscientific Age," John Danz Lecture Series, 1963

Universal education is probably a good thing, but you could teach good as well as bad — you can teach falsehood as well as truth. The communication between nations as it develops through a technical development of science should certainly improve the relations between nations. Well, it depends what you communicate. You can communicate truth and you can communicate lies. You can communicate threats or kindness.

– Galileo Symposium, "What Is and What Should Be the Role of Scientific Culture in Modern Society," September 1964

My theory is that loss of common interest — between the engineers and scientists on the one hand and management on the other — is the cause of the deterioration in cooperation.

– "Mr. Feynman Goes to Washington," 1987

I expect to stay away from the President's Board of Master Plumbers as much as possible, and I hope you do not draft me.

– Letter to Congressman Barber B. Conable Jr., November 1965

You write that true Americans have a big generous heart, which shows only what a big and generous heart you have. For you must know that a great nation, at least one where the British ideas of freedom flourish, is very complex, and side by side lie the great and the mean, the generous and the selfish, just as they lie side by side in each man.

– Letter to Reverend John Alex and Mrs. Marjorie Howard, December 1965 (*Perfectly Reasonable Deviations from the Beaten Track*, p. 184)

No government has the right to decide on the truth of scientific principles, nor to prescribe in any way the character of the questions investigated. Neither may a government determine the aesthetic value of artistic creations, nor limit the forms of literary or artistic expression. Nor should it pronounce on the validity of economic, historic, religious, or philosophical doctrines. Instead it has a duty to its citizens to maintain the freedom, to let those citizens contribute to further adventure and the development of the human race.

– "The Uncertainty of Values," John Danz Lecture Series, 1963

I decided at the time, I believe, that the morally right thing to do was to protect ourselves; I felt there was a great evil around, and that this evil would only grow if it had more technical power. The only way that I knew how to prevent that was to get there earlier so that we could prevent them from doing it, or defeat them.

– BBC, "The Pleasure of Finding Things Out," 1981

Mr. Hollingsworth shows nerve in calling me an AEC specialist. Is every scientist in the country working for the Atomic Energy Commission?

– Letter to Lewis L. Strauss, July 6, 1956

[On the Rogers Commission:] I have a unique qualification — I am completely free and there are no levers that can be used to influence me and I am reasonably straightforward and honest. There are exceedingly powerful political forces and consequences involved here. But although people have explained that to me from different points of view I disregarded them all, and proceed with apparently naive and single-minded purpose to the one end: first, why, physically, the shuttle failed, leaving later the question of why humans made, apparently, bad decisions when they did.

– Letter to Gweneth and Michelle Feynman, February 12, 1986
(*Perfectly Reasonable Deviations from the Beaten Track*, p. 398)

[After a UN meeting on atomic energy:] Ever since then, I've had a much better idea how government works, and what the hell's the matter with it. I mean, that those things that are vital to be decided are decided too easily. I mean, it's great that a man can decide so quickly. So can a die decide quickly? It's just — it's bad. It was a very serious thing.

– Interview with Charles Weiner, June 27, 1966 (Niels Bohr Library and Archives with the Center for the History of Physics)

There are an awful lot of light decisions of historical importance.

– Interview with Charles Weiner, June 27, 1966 (Niels Bohr Library and Archives with the Center for the History of Physics)

[On company workers:] They have a conflict, and tensions, and usually they either give up completely, or there's another tension that happens. If they get 100 percent involved in the company work, then another trouble happens. There's always some guy on top of them, the boss, who is dumber technically than they are, and who makes the decision as to what they ought to do, and whether what they do is worthwhile.

– Interview with Charles Weiner, June 28, 1966 (Niels Bohr
   Library and Archives with the Center for the History of Physics)

The real question of government versus private enterprise is argued on too philosophical and abstract a basis. Theoretically, planning may be good. But nobody has ever figured out the cause of government stupidity — and until they do (and find the cure), all ideal plans will fall into quicksand.

– *What Do You Care What Other People Think?*, pp. 90–91

So, very delicately, amongst all these liberal-minded scientific guys agreeing to such a proposition, we finally got the censorship set up.

– UCSB talk, "Los Alamos from Below," February 1975

The other thing that gives a scientific man the creeps in the world today are the methods of choosing leaders — in every nation. Today, for example, in the United States, the two political parties have decided to employ public relations men, that is, advertising men, who are trained in the necessary methods of telling the truth and lying in order to develop a product. This wasn't the original idea. They are supposed to discuss situations and not just make up slogans. It's true, if you look in history, however, that choosing political leaders in the United States has been on many different occasions based on slogans.

– Galileo Symposium, "What Is and What Should Be the Role of Scientific Culture in Modern Society," September 1964

Doubt and Uncertainty

But I don't have to know an answer. I don't feel frightened by not knowing things, by being lost in a mysterious universe without any purpose, which is the way it really is, so far as I can tell. It doesn't frighten me.

– BBC, "The Pleasure of Finding Things Out," 1981

[Doubt] is not a new idea; this is the idea of the age of reason. This is the philosophy that guided the men who made the democracy that we live under. The idea that no one really knew how to run a government led to the idea that we should arrange a system by which new ideas could be developed, tried out, tossed out, more new ideas brought in; a trial and error system. This method was a result of the fact that science was already showing itself to be a successful venture at the end of the 18th century. Even then it was clear to socially minded people that the openness of the possibilities was an opportunity, and that doubt and discussion were essential to progress into the unknown.

– "The Value of Science," December 1955

I can live with doubt, and uncertainty, and not knowing. I think it's much more interesting to live not knowing anything than to have answers which might be wrong. I have approximate answers, and possible beliefs, and different degrees of certainty about different things, but I'm not absolutely sure of anything.

– BBC, "The Pleasure of Finding Things Out," 1981

American ideals of freedom are the same as the ideals of scientific development. The ultimate development of the potentials of mankind requires an allowance for error.

– Notes for talk on "Science in America"

Now, we scientists are used to this, and we take it for granted that it is perfectly consistent to be unsure, that it is possible to live and *not* know. But I don't know whether everyone realizes this is true. Our freedom to doubt was born out of a struggle against authority in the early days of science. It was a very deep and strong struggle: permit us to question — to doubt — to not be sure. I think that it is important that we do not forget this struggle and thus perhaps lose what we have gained.

– "The Value of Science," December 1955

This business about not understanding is a very serious one that we have between scientists and an audience. And I want to work with you because I want to tell you something: Students do not understand it either, and that's because the professor doesn't understand it, which not a joke but very interesting.

– "QED: Photons — Corpuscles of Light," Sir Douglas Robb Lectures, University of Auckland, June 1979

People who hear that all I'm going to do is make a couple of arrows on a board to calculate the chance that something happens — say that this guy doesn't know physics. But this is the guy that knows that that's what you have to do and admits, therefore, that he doesn't know why he's doing what he does. And you can have the confidence that when I say I don't know what I'm doing, probably no one else does either.

– "QED: Fits of Reflection and Transmission," Sir Douglas Robb Lectures, University of Auckland, 1979

Everything has to be checked out very carefully. Otherwise you become one of those people who believe all kinds of crazy stuff and doesn't understand the world they're in. Nobody understands the world they're in, but some people are better off at it than others.

– "This Unscientific Age," John Danz Lecture Series, 1963

It is in the admission of ignorance and the admission of uncertainty that there is a hope for the continuous motion of human beings in some direction that doesn't get confined, permanently blocked, as it has so many times before in various periods in the history of man.

– "The Uncertainty of Values," John Danz Lecture Series, 1963

It would have been unscientific not to guess.

– "The Uncertainty of Science," John Danz Lecture Series, 1963

Does this mean that physics, a science of great exactitude, has been reduced to calculating only the probability of an event, and not predicting exactly what will happen? Yes.

– *QED: The Strange Theory of Light and Matter*, p. 19

I think it's much more interesting to live not knowing than to have answers which might be wrong. I have approximate answers and possible beliefs and different degrees of uncertainty about different things, but I am not absolutely sure of anything and there are many things I don't know anything about, such as whether it means anything to ask why we're here. I don't have to know an answer. I don't feel frightened not knowing things, but being lost in a mysterious universe without any purpose, which is the way it really is, as far as I can tell.

– BBC, "The Pleasure of Finding Things Out," 1981

I think that when we know that we actually do live in uncertainty, then we ought to admit it; it is of great value to realize that we do not know the answers to different questions. This attitude of mind — this attitude of uncertainty — is vital to the scientist, and it is this attitude of mind which the student must first acquire. It becomes a habit of thought. Once acquired, one cannot retreat from it anymore.

– Galileo Symposium, "What Is and What Should Be the Role of Scientific Culture in Modern Society," September 1964

I've learned how to live without knowing. I don't have to be sure I'm succeeding, as I said before about science, I think my life is fuller because I realize that I don't know what I'm doing. I'm delighted with the width of the world!

– *Omni* interview, February 1979

[On "typewriter science":] Some intellectual sits at the typewriter and writes it all out as if the information were really known. The intellectual never says, "I don't know this," or "I'm not really sure." If he were to do so, he couldn't sell his articles because somebody else would come along and say that they have all the answers.

– *U.S. News and World Report* interview, February 1985

All scientific knowledge is uncertain. This experience with doubt and uncertainty is important. I believe that it is of very great value, and one that extends beyond the sciences. I believe that to solve any problem that has never been solved before, you have to leave the door to the unknown ajar. You have to permit the possibility that you do not have it exactly right. Otherwise, if you have made up your mind already, you might not solve it.

– "The Uncertainty of Science," John Danz Lecture Series, 1963

So what we call scientific knowledge today is a body of statements of varying degrees of certainty. Some of them are most unsure; some of them are nearly sure; but none is absolutely certain.

– "The Uncertainty of Science," John Danz Lecture Series, 1963

It is not possible to predict exactly what'll happen in any circumstance.

– Audio recording of Feynman Lectures on Physics, Lecture 2, September 29, 1961

The thing you can't face is not knowing what somebody's cooking up. That's hard, not knowing. But any real thing, you just sit there, take it for reality, and see what you do under the circumstances.

– Interview with Charles Weiner, March 5, 1966 (Niels Bohr Library and Archives with the Center for the History of Physics)

There is no harm in being uncertain. It is better to say something and not be sure than to not say anything at all.

– "The Uncertainty of Science," John Danz Lecture Series, 1963

It is of paramount importance, in order to make progress, that we recognize this ignorance and this doubt. Because we have the doubt, then we propose looking in new directions for new ideas.

– "The Uncertainty of Science," John Danz Lecture Series, 1963

If we were not able or did not desire to look in any new direction, if we did not have a doubt or recognize ignorance, we would not get any new ideas.

– "The Uncertainty of Science," John Danz Lecture Series, 1963

What we call scientific knowledge today is a body of statements of varying degrees of certainty.

– "The Uncertainty of Science," John Danz Lecture Series, 1963

I feel a responsibility as a scientist who knows the great value of a satisfactory philosophy of ignorance, and the progress made possible by such a philosophy, progress which is the fruit of freedom of thought. I feel a responsibility to proclaim the value of this freedom and to teach that doubt is not to be feared, but that it is to be welcomed as the possibility of a new potential for human beings.

– "The Uncertainty of Science," John Danz Lecture Series, 1963

I want to maintain here that in the admission of ignorance and the admission of uncertainty that there is a hope for the continuous motion of human beings in some direction that doesn't get confined, permanently blocked as it has so many times before in various periods in the history of man.

– "The Uncertainty of Values," John Danz Lecture Series, 1963

The only way that we will make a mistake is that in the impetuous youth of humanity we will decide we know the answer.

– "The Uncertainty of Values," John Danz Lecture Series, 1963

All my life I have been inspired by Galileo and his struggle with the Church, his struggle for a freedom to doubt, and the struggle that went along with other people connected with Galileo.

– Galileo Symposium, "What Is and What Should Be the Role of Scientific Culture in Modern Society," September 1964

One way of looking at the uncertainty in the world is supposing that these things are there but that we only see them crudely. This idea we have to get rid of it doesn't work!

– Esalen lecture, "Quantum Mechanical View of Reality (Part 1)," October 1984

The thing that's unusual about scientists is that while they're doing whatever they're doing, they're not so sure of themselves as others usually are. They can live with steady doubt, think "maybe it's so" and act on that, all the time knowing it's only "maybe."

– *Omni* interview, February 1979

I believe that a scientist looking at nonscientific problems is just as dumb as the next guy and when he talks about a nonscientific matter, he will sound as naive as anyone untrained in the matter.

– "The Value of Science," December 1955

When we are talking to each other at these high and complicated levels, and we think we are speaking very well, that we are communicating, but what we are really doing is having a some kind of big translation scheme going on, translating what this fellow says into our images, which are very different.

– BBC, "Fun to Imagine" television series, 1983

There is no harm in doubt and skepticism, for it is through these that new discoveries are made.

– Letter to Armando Garcia, December 1985 (*Perfectly Reasonable Deviations from the Beaten Track*, p. 396)

In physics the truth is rarely perfectly clear, and that is certainly universally the case in human affairs. Hence, what is not surrounded by uncertainty cannot be the truth.

– Letter to the editor of the California Tech, February 1976

You may think that it might be possible to invent a metaphysical system for religion which will state things in such a way that science will never find itself in disagreement. But I do not think that it is possible to take an adventurous and ever-expanding science that is going into an unknown, and to tell the answer to questions ahead of time and not expect that sooner or later, ne matter what you do, you will find that some answers of this kind are wrong.

> – "The Uncertainty of Values," John Danz Lecture Series, 1963
> (*The Meaning of It All,* pp. 46–47)

Science is the belief in the ignorance of experts.

> – National Science Teachers Association Fourteenth Convention
> lecture, "What Is Science?" April 1966

When you're thinking about something you don't understand, you have a terrible, uncomfortable feeling called confusion. It's a very difficult and unhappy business. And so most of the time you're rather unhappy, actually, with this confusion. You can't penetrate this thing. Now, is the confusion because we're all some kind of apes that are kind of stupid, trying to figure out how to put the sticks together to reach the banana and we can't quite make it, the idea. And I get that feeling all the time, that I'm an ape trying to put the two sticks together. So I always feel stupid. Once in a while, though, the sticks go together on me and I reach the banana.

> – Swedish television interview on Nobel Prize winners, 1965

To make progress in understanding, we must remain modest and allow that we do not know. Nothing is certain or proved beyond all doubt. You investigate for curiosity, because it is unknown, not because you know the answer. And as you develop more information in the sciences, it is not that you are finding out the truth, but that you are finding out that this or that is more or less likely.

> – "The Relation of Science and Religion," May 1956

The fact that you are not sure means that it is possible that there is another way someday.

– "The Uncertainty of Values," John Danz Lecture Series, 1963

We absolutely must leave room for doubt or there is not progress and there is no learning. There is no learning without having to pose a question. And a question requires doubt. People search for certainty. But there is no certainty. People are terrified — how can you live and not know? It is not odd at all. You only think you know, as a matter of fact. And most of your actions are based on incomplete knowledge and you really don't know what it is all about, or what the purpose of the world is, or know a great deal or other things. It is possible to live and not know.

– "The Role of Scientific Culture in Modern Society"

Education and Teaching

I don't believe I can really do without teaching. The reason is, I have to have something so that when I don't have any ideas and I'm not getting anywhere I can say to myself, "At least I'm living; at least I'm doing something; I'm making some contribution" — it's just psychological.

– *Surely You're Joking, Mr. Feynman!*, p. 165

You'll learn infinitely better and easier and more completely by picking a problem for yourself that you find interesting to fiddle around with some kind of thing that you heard that you don't understand, or you want to analyze further, or want to do some kind of a trick with — that's the best way to learn something.

– *Feynman's Tips on Physics*, p. 15

In teaching science you are telling new minds about the most wondrous world they live in, you are passing on the greatest ideas man has ever contemplated: the fathomless space, the infinite gyrations of the atoms, the web of interaction of all life, and of life with the inanimate. And this vast sea of knowledge is surrounded on all sides by much vaster areas of ignorance, which are not things to fear but challenges to learn about.

– Notes for talk on "Science in America"

In fact, it's impossible in the long run to do everything by memory. That doesn't mean to do nothing by memory — the more you remember, in a certain sense, the better it is — but you should be able to recreate anything that you forgot.

– *Feynman's Tips on Physics*, p. 41

This is, as a matter of fact, the way you start on any complicated or unfamiliar problem: You first get a rough idea; then you go back when you understand it better and do it more carefully.

– *Feynman's Tips on Physics*, p. 77

When it came time for me to give my talk on the subject, I started off by drawing an outline of the cat and began to name the various muscles. The other students in the class interrupt me. "We know all that!" "Oh," I say, "you do? Then no wonder I can catch up with you so fast after you've had four years of biology." They had wasted all their time memorizing stuff like that, when it could be looked up in fifteen minutes.

– *Surely You're Joking, Mr. Feynman!*, p. 72

I often remember vividly my most enjoyable trip. Many things bring it to mind — like the dark blue T-shirt in my drawer at home — or the interference picture in my office — or just now when my secretary asked me if I wanted to talk to students at a nearby university (USC 20 miles) or high school. My answer was that I will talk to students anytime they are near enough to home, or are at Vancouver, B.C.

– Letter to Mariela Johansen, April 1975

The students may not be able to see the thing I want to answer, or the subtleties I want to think about, but they remind me of a problem by asking questions in the neighborhood of that problem. It's not so easy to remind yourself of these things.

– *Surely You're Joking, Mr. Feynman!*, p. 166

Don't look up the answer; just figure it out. After all, it's only nature; she can't beat you. If you think hard enough, you'll figure it out.

– Audio recording of Feynman Lectures on Physics, Lecture 30, February 20, 1962

Learn by trying to understand simple things in terms of other ideas — always honestly and directly. What keeps the clouds up, why can't I see stars in the daytime, why do colors appear on oily water, what makes the lines on the surface of water being poured from a pitcher, why does a hanging lamp swing back and forth — and all the innumerable little things you see all around you. Then when you have learned to explain simpler things, so you have learned what an explanation really is, you can then go on to more subtle questions.

– Letter to Master Ashok Arora, January 1967 (*Perfectly Reasonable Deviations from the Beaten Track*, p. 230)

It has nothing to do with what you're supposed to learn. All it is is that you can learn something, and it's just useful to learn as much as possible. However, it's imperative to learn a certain minimum amount in order to go on and learn something else.

– Audio recording of Feynman Lectures on Physics, Lecture 17, November 28, 1961

I am going to try and tell what the world of light and electrons looks like from the point of view of modern physics. This is quite an order, and I may fail badly — but let us try.

– On his lecture topic for the Sir Douglas Robb Lectures, June 1979

Those are just names, and people use them, so we'll use them.

– Audio recording of Feynman Lectures on Physics, Lecture 30, February 20, 1962

Foreign languages have been taught all over for so long! Are there not studies of methods and results — inconclusive studies at least that show that any method is worse than any other that can be used to prove or disprove anything? That is just what you need to generate doubt and humbleness and an atmosphere more conducive to uncertainty.

– Letter to Dr. Amos J. Lessard, February 1983

If you understand something, you can remember it, when you work it out yourself.

– Interview with Charles Weiner, March 5, 1966 (Niels Bohr Library and Archives with the Center for the History of Physics)

As I get more experience I realize that I know nothing whatsoever as to how to teach children arithmetic. I did some write some things before I reached my present state of wisdom.

– Letter to Beryl S. Cochran, April 1967 (*Perfectly Reasonable Deviations from the Beaten Track*, p. 241)

I don't know anything about small children. I have one, so I know that I don't know.

– National Science Teachers Association Fourteenth Convention lecture, "What Is Science?" April 1966

You'll see I'll never get anywhere with the physics if I keep talking about all this stuff. But it's interesting. So what? So it doesn't make any difference. So you learn something else.

– Audio recording of Feynman Lectures on Physics, Lecture 14, November 14, 1961

It is necessary to learn words. It is not a science. That doesn't mean just because it is not a science that we don't have to teach the words. We are not talking about what to teach; we are talking about what science is. It is not science to know how to change centigrade to Fahrenheit. It's necessary, but it is not exactly science. In the same sense, if you were discussing what art is, you wouldn't say art is the knowledge of the fact that a 3-B pencil is softer than a 2-H pencil. It's a distinct difference. That doesn't mean an art teacher shouldn't teach that, or that an artist gets along very well if he doesn't know that.

– National Science Teachers Association Fourteenth Convention lecture, "What Is Science?" April 1966

It's natural to explain an idea in terms of what you already have in your head. Concepts are piled on top of each other: This idea is taught in terms of that idea, and that idea is taught in terms of another idea, which comes from counting, which can be so different for different people!

– *What Do You Care What Other People Think?*, p. 59

Interest is an emotion — like love. It is not a property of a subject.

– Notes

In any thinking process there are moments when everything is going good and you've got wonderful ideas. Teaching is an interruption, and so it's the greatest pain in the neck in the world.

– *Surely You're Joking, Mr. Feynman!*, p. 165

In psychology there is a profound question of what type of facility is it that permits a child to learn a language from simply hearing it spoken and seeing it used. We are far from knowing how it is done. It is even very hard to see how it can be done. But it is done by every child. We cannot expect to solve such problems by studying machines. Nevertheless, it is an intriguing academic investigation to see, at least in principle, some way that it might be done by a machine.

– Letter to R. B. Leighton, April 1974

You ask, how can this guy teach, how can he be motivated if he doesn't know what he's doing. As a matter of fact, I love to teach. I like to think of new ways of looking at things as I explain them, to make them clearer — but maybe I'm not making them clearer. Probably what I'm doing is entertaining myself.

– *Omni* interview, February 1979

One does not learn a subject by using the words that people who know the subject use in discussing it. One must learn how to handle the ideas and then, when the subtleties arise which require special language, that special language can be used and developed easily. In the meantime, clarity is the desire.

– "New Mathematics," written for the California State Department of Education

They treated me so nicely in Vancouver that now I now the secret of how to really be entertained and give talks: Wait for the students to ask you.

– *Surely You're Joking, Mr. Feynman!*, p. 303

I don't know what's the matter with people: they don't learn by understanding; they learn by some other way — by rote or something. Their knowledge is so fragile.

– *Surely You're Joking, Mr. Feynman!*, pp. 36–37

I am a successful lecturer in physics for popular audiences. The real entertainment gimmick is the excitement, drama and mystery of the subject matter. People love to learn something, they are "entertained" enormously by being allowed to understand a little bit of something they never understood before. One must have faith in the subject and in people's interest in it.

– Letter to Ralph Bown, March 1958, *Perfectly Reasonable Deviations from the Beaten Track*, p. 98

I believe that all of the exercises in all of the books, from the first to the eighth year, ought to be understandable to any ordinary adult — that is, the question that one is trying to find out should be clear to every person.

– "New Mathematics," written for the California State Department of Education

Physics has a profound effect on all the other sciences. So that students from many other sciences find themselves studying it.

– Audio recording of Feynman Lectures on Physics, Lecture 3, October 3, 1961

See, with all these things the big trouble is the words. They sound like something terrible, but the idea isn't really so bad.

– Audio recording of Feynman Lectures on Physics, Lecture 8, October 20, 1961

That's just symbols — you have to know the idea.

– Audio recording of Feynman Lectures on Physics, Lecture 8, October 20, 1961

Don't copy it down; just listen to what it says. Otherwise you don't understand anything when it's finished.

– Audio recording of Feynman Lectures on Physics, Lecture 15, November 17, 1961

Nobody knows how to teach physics, or to educate people — that's a fact, and if you don't like the way it's being done, that's perfectly natural. It's impossible to teach satisfactorily: For hundreds of years, even more, people have been trying to figure out how to teach, and nobody has ever figured it out.

– *Feynman's Tips on Physics*, p. 15

I suffer from the disease that all professors suffer from — that is there never seems to be enough time, and I invented more problems than undoubtedly we'll be able to do, and therefore I've tried to speed things up by writing some things on the board beforehand with the illusion that every professor has: that if he talks about more things, he'll teach more things. Of course, there's only a finite rate at which material can be absorbed by the human mind, yet we disregard that phenomenon, and in spite of it we go too fast.

– *Feynman's Tips on Physics*, p. 71

One of the things one learns in school which is incorrect is that problems are relatively easy, if they're formed, you can set 'em up, you can solve them — which isn't at all true.

– Interview with Charles Weiner, March 5, 1966 (Niels Bohr Library and Archives with the Center for the History of Physics)

I thought that one of the troubles with all the courses in physics was that they just said: You learn all this, you learn all that, and when you come out the other end you'll understand the connections. But there's no map, "guide to the perplexes," you see. So I want to make a map. But it turns out it's not a feasible design. I mean, I just never made such a map.

– Interview with Charles Weiner, June 28, 1966 (Niels Bohr Library and Archives with the Center for the History of Physics)

But there must be people living who aren't listening to the lectures of some professor, who are sitting just reading the book and thinking for themselves. They must get something out of it. So f I keep some hope that that's worth something to them, maybe I can feel better about the whole thing.

– Interview with Charles Weiner, June 28, 1966 (Niels Bohr Library and Archives with the Center for the History of Physics)

What I really was doing was teaching myself. I wasn't interested in publishing at all. But I did discover a lot of things. You see, I thought everybody else knew all these things. In the meantime I was trying to teach myself. So I learned a lot of things that weren't known, or a few things that weren't well known. And I checked — things that people have noticed later as being simple, sometimes I noticed a little ahead. But the main thing I was doing was teaching myself.

– Interview with Charles Weiner, February 4, 1973 (Niels Bohr
  Library and Archives with the Center for the History of Physics)

I finally figured out a way to test whether you have taught an idea or you have only taught a definition. Test it this way: You say, "Without using the new word which you have just learned, try to rephrase what you have just learned in your own language."

– National Science Teachers Association Fourteenth Convention
  lecture, "What Is Science?" April 1966

To learn a mystic formula for answering questions is very bad.

– National Science Teachers Association Fourteenth Convention
  lecture, "What Is Science?" April 1966

A child should be given a child's answer. "Open it up; let's look at it."

– National Science Teachers Association Fourteenth Convention
  lecture, "What Is Science?" April 1966

I have your letter of February 6, requesting my views on teaching science to youngsters. I have just one suggestion — never mind a the big talk about materials — get a good teacher and back her up. There's no other way!

- Letter to Douglas O'Brien (Sunset Hill Elementary School), March 1967

I don't wish to spare the time on the non-science students. Anything I say would be either easily available by reading or talking to the science students, or else it would be over their heads.

- Letter to Franklin W. Stahl (University of Oregon), April 1961

(In response to a child's question, "Does time exist?") Suppose it didn't. Then what?

- From notes for "About Time" program, 1957

When we come to consider the words and definitions which children ought to learn, we should be careful not to teach "just" words. It is possible to give an illusion of knowledge by teaching the technical words which someone uses in a field which sound unusual to ordinary ears, without at the same time teaching some ideas or some facts using these words which require the special words in the special way and with the special care that they have been defined.

- "New Mathematics," written for the California State Department of Education, 1965

This ends the main points that I wanted to make — the first that there must be a freedom of thought; second, that we do not want to teach just words, and third, that subjects should not be introduced without explaining the purpose or reason or without producing any yield as the subject is studied during the first eight years of school.

– Unpublished extended version of "New Mathematics," written for the California State Department of Education, 1965

Students have a resilience and a skill at recognizing "all that jazz" as just "jazz." It was a child who understood the emperor's clothes!

– Letter to Richard Godshall, March 1966 (*Perfectly Reasonable Deviations from the Beaten Track*, p. 218)

I believe that a book should be only an assistance to a good teacher and not a dictator.

– Letter to Richard Godshall, March 1966 (*Perfectly Reasonable Deviations from the Beaten Track*, p. 218)

Stay human and on your pupil's side.

– Letter to Richard Godshall, March 1966 (*Perfectly Reasonable Deviations from the Beaten Track*, p. 218)

It would be very good for the students to learn that science is an ongoing subject with new research always modifying old ideas and I myself would be happy to see the careful criticism of my ideas that your physicists have made.

– Letter to Sandor Solt, April 1969 (*Perfectly Reasonable Deviations from the Beaten Track*, p. 251)

Simple questions with complicated answers are always asked by dull students. Only intelligent students have been trained to ask complicated questions with simple answers — as any teacher knows (and only teachers think there are any simple questions with simple answers).

– Letter to Professor Michael H. Hart, December 1980 (*Perfectly Reasonable Deviations from the Beaten Track*, p. 330)

I have much experience only in teaching graduate students in physics, and as a result of that, I know that I don't know how to teach.

– National Science Teachers Association Fourteenth Convention lecture, "What Is Science?" April 1966

We decided that we don't have to teach a course in elementary quantum mechanics in the graduate school any more. When I was a student, they didn't even have a course in quantum mechanics in the graduate school, it was considered too difficult a subject. When I first started to teach, we had one. Now we teach it to undergraduates. We discover now that we don't have to have elementary quantum mechanics for graduates from other schools. Why is it getting pushed down? Because we are able to teach better in the university, and that is because the students coming up are better trained.

– National Science Teachers Association Fourteenth Convention lecture, "What Is Science?" April 1966

think it is very important — at least it was to me — that if you are going to teach people to make observations, you should show that something wonderful can come from them.

– National Science Teachers Association Fourteenth Convention lecture, "What Is Science?" April 1966

I now know that it's possible for a night's work of a graduate student to be done in ten seconds by a professor.

– CERN talk, December 1965

There are many new plans in many countries for trying to teach physics, which shows that nobody is satisfied with any method. It is likely that many of the new plans look good, for nobody has tried them long enough to find out what is the matter with them; whereas all the old methods have been with us long enough to show their faults clearly.

– "The Problem of Teaching Physics in Latin America," 1963

The fact is that nobody knows very well how to tell anybody else how to teach. So when we try to figure out how to teach physics we must be somewhat modest, because nobody really knows how. It is at the same time a serious problem and an opportunity for new discoveries.

– "The Problem of Teaching Physics in Latin America," 1963

Science is an activity of men; to many men, it is a great pleasure and it should not be denied to the people of a large part of the world simply because of a fault or lack in the educational system

– "The Problem of Teaching Physics in Latin America," 1963

From memorizing, knowledge is not understood, and the beauty of nature is not appreciated. It does not tell how things were found out, or reveal the value of an inventive free mind.

– "The Problem of Teaching Physics in Latin America," 1963

I understand I am supposed to make some response and so I planned a talk on my philosophy of teaching. After pondering it at great length, I discovered I had nothing but clichés and unimportant things, to say, and so I called up to ask if I could talk on something else, and so I would like to talk on something in physics itself rather than to talk about teaching because I don't know anything about teaching.

– Oersted Medal acceptance speech, 1972

Say you've got a disease, Wener's granulomatosis or whatever, and you look it up in a medical reference book. You may well find that you then know more about it than your doctor does, although he spent all that time in medical school, you see? It's much easier to learn about some special, restricted topic than a whole field.

– *Omni* interview, February 1979

I'm disappointed with my students all the time. I'm not a teacher who knows what he's doing.

– *Omni* interview, February 1979

My task is to explain all this and convince you not to turn away because it appears incomprehensible. That's what it takes four years for us to do to a student: to get him so he doesn't run away because it looks crazy. The thing that's exciting about this is that Nature is strange as it can be in this sense!

– "QED: Photons — Corpuscles of Light," Sir Douglas Robb Lectures, University of Auckland, June 1979

I sometimes feel that it would be much better not to educate our children in such subjects as mathematics and science. If we left youngsters alone, there would be a better chance that, by accident, the kids would find a good book or an old textbook or a television program that would excite them. But when youngsters go to school, they learn that these subjects are dull, horrible, and impossible to understand.

– *U.S. News and World Report* interview, February 1985

[On school textbooks:] They don't try to make subjects easier to understand. They try to make it easier to know what to do to pass the test and please the teacher.

– *U.S. News and World Report* interview, February 1985

One has the advantage of using questions to finish the lecture.

– "QED: Fits of Reflection and Transmission," Sir Douglas Robb Lectures, University of Auckland, 1979

My theory is that an excited and enthusiastic teacher teaching in a new and experimental situation — trying new things out — exudes so much in personality and energy that the students (or some students, at least) cannot help but respond. They respond magnificently.

– Letter to Mr. Robert Bonic, January 1974

General education is required for such a large number of students that a very large number of teachers are needed. Because there are so many teachers and because only a few people are really outstanding in anything, we must realize that most teachers must be either mediocre or dull. This is not a criticism of the profession — it is merely a matter of arithmetic.

– Letter to Mr. Robert Bonic, January 1974

It is too easy to memorize a lot or read a lot and not really understand anything.

– Letter to Apostolos Tournas, February 1985

About your attempts at teaching: It is so difficult isn't it — you can lead the horses to water but you can't make the damn fools drink!

– Letter to Malcolm Joseph, January 1982

One of my little pleasures in life is to go to the Van Nuys High School once or twice a year to answer questions for the science students of Mr. Coutts' classes. This activity was initiated many years ago by Mr. Coutts, and I look forward to doing it every year. The questions are on anything: relativity, black holes, clouds, spinning tops, magnetic force, you name it. The class is alive and very interested and seems to enjoy it as much as I.

– Letter to Ms. Melinda Jan, April 1985 (*Perfectly Reasonable Deviations from the Beaten Track*, p. 380)

Advice to a student:] If your professors and fellow students seem to know some things but seem to be oblivious to other things, that does not exclude you from learning what they know whilst remaining deeply aware of what they are blind to.

– Letter to Mr. Alan Woodward, March 1982 (*Perfectly Reasonable Deviations from the Beaten Track*, p. 345)

Most theorists don't know that wire space is a problem. For them a wire is an idealized thin string that doesn't take up any space, but real computer designers soon discover that they can't get enough wires in.

  – "The Remarkable Dr. Feynman," *Los Angeles Times Magazine*,
    April 20, 1986

When the student doesn't understand the professor, usually the student thinks it's because he's dopey and didn't catch on. This time it was because the professor wasn't saying anything that was sensible.

  – "Tiny Computers Obeying Quantum Mechanical Laws,"
    *New Directions in Physics: The Los Alamos 40th Anniversary
    Volume*, 1987

Easy to teach that any subject is dull — very rare but not impossible to teach a subject is interesting.

  – Notes

Biology is not simply writing information; it is doing something about it.

  – "There's Plenty of Room at the Bottom," December 1959

If you look closely enough at anything, you will see that there i nothing more exciting than the truth, the pay dirt of the scientist

  – "The Uncertainty of Science," John Danz Lecture Series, 1963

You teach the value of a free, inquiring, and discovering mind — the kind of mind that built America and that America was built fo

  – Notes for talk on "Science in America"

There may be some idea that's difficult to understand the first time you study it. For example, Einstein's theory or something like this. And a man trying to learn it can't understand it. Later he finally understands it — say, when he goes to teach it, he finally understands it. He thinks that his particular way of understanding it is a very much clearer than the way it was presented to him before. Therefore, big deal, he puts out a paper — new way of looking at it! Actually, it's not a new way. I mean, maybe it is a little bit new, but it's very personal, and it's not sufficiently different.

– Interview with Charles Weiner, March 5, 1966 (Niels Bohr Library and Archives with the Center for the History of Physics)

I worked hard at teaching in the beginning, and they were OK. I think the students were satisfied. Later, you give the same course over again, and you don't work so hard if you don't reorganize it. I got more and more careless about teaching. And if I teach something I've taught before, it's not any longer a good course, because I borrow so much stuff from before; and I'm so lazy about correcting papers and preparing the courses, I don't think they're any good any more. I think I'm getting less and less careful as a teacher, relatively. I mean, I'm still useful, but I think I used to be good, really good, relatively. And now I'm lazy.

– Interview with Charles Weiner, June 27, 1966 (Niels Bohr Library and Archives with the Center for the History of Physics)

There's an awful lot of rote learning, and a lot of mistaking knowledge for the right technical words and so on, you know. A guy who says the right words is thought to know something. I didn't bother but I could have taught my child, after he learned to talk, to say — and I thought I would, just for the fun of it, to demonstrate this, but I didn't bother the poor boy — but it's not at all impossible to teach a child to say that pi is the ratio of the circumference to the diameter of a circle. It's just as easy to teach him that as to teach him a nursery rhyme. And then to say that pi is numerically equal to 3.14159. That way you can get fooled. You haven't the slightest idea what you're talking about, and you sound just fine.

- Interview with Charles Weiner, June 28, 1966 (Niels Bohr Library and Archives with the Center for the History of Physics)

I never know when I'm lecturing or teaching what the response is.

- Interview with Charles Weiner, June 28, 1966 (Niels Bohr Library and Archives with the Center for the History of Physics)

Teach them the culture of our government and our times, and prepare the better of them to be able to watch and appreciate or perhaps to take part in the greatest adventure that the human mind has ever embarked on.

- Notes for talk on "Science in America"

Today we do not have the power of expression to tell a student how to understand physics physically. We can write the laws, but we still can't say how to understand them physically. The only way you can understand physics physically, because of our lack of machinery for expressing this, is to follow the dull, Babylonian way of doing a whole lot of problems until you get the idea. That's all I can do for you. And the students who didn't get the idea in Babylonia flunked, and the guys who did get the idea died, so it's all the same!

– *Feynman's Tips on Physics*, p. 50

If we're going to explain this theory, the first question is will you understand it? Will you understand the theory? When I tell you first that the first time we really fully explain it to our physics students, they're in the third year graduate physics, then you think the answer's going to be no. And that's correct: You will not understand it.

– "QED: Photons — Corpuscles of Light," Sir Douglas Robb Lectures, University of Auckland, June 1979

I think we should teach them that the purpose of knowledge is to appreciate wonders even more. And that the knowledge is just to put into correct framework the wonder that nature is.

– On teaching science to the public, Galileo Symposium, "What Is and What Should Be the Role of Scientific Culture in Modern Society," September 1964

If the professors of English will complain to me that the students who come to the universities, after all those years of study, still cannot spell "friend," I say to them that something's the matter with the way you spell friend.

– "This Unscientific Age," John Danz Lecture Series, 1963

Advice and Inspiration

If you have any talent, or any occupation that delights you, do it, and do it to the hilt. Don't ask why or what difficulties you may get into.

– Letter to student Frederich Hipp, April 1961 (*Perfectly Reasonable Deviations from the Beaten Track*, p. 120)

You have no responsibility to live up to what other people think you ought to accomplish. I have no responsibility to be like they expect me to be. It's their mistake, not my failing.

– *Surely You're Joking, Mr. Feynman!*, p. 172

You ask me if an ordinary person could ever get to be able to imagine these things like I imagine them. Of course! I was an ordinary person who studied hard. There are no miracle people. It just happens they get interested in this thing and learn all this stuff, but they're just people.

– BBC, "Fun to Imagine" television series, 1983

[On revelation:] The hope of that gold — that can keep you going.

– Yorkshire Television program, "Take the World from Another Point of View," 1972

Know how to solve every problem that has been solved.

– Written on blackboard when he died

Just because Feynman says he is pro-nuclear power, isn't any argument at all worth paying attention to because I can tell you (for I know) that Feynman really doesn't know what he is talking about when he speaks of such things. He knows about other things (maybe). Don't pay attention to "authorities," think for yourself.

– Letter to student Mark Minguillon, August 1976 (*Perfectly Reasonable Deviations from the Beaten Track*, p. 305)

Any real knowledge has to have been found out somehow. If an expert tells you that "a great man invented it" and the ideas cannot be explained, then be suspicious.

– *U.S. News and World Report* interview, February 1985

All you have to do is, from time to time — in spite of everything, just try to examine a problem in a novel way. You won't "stifle the creative process" if you remember to think from time to time. Don't you have time to think?

– Letter to Michael E. Stanley, March 1975 (*Perfectly Reasonable Deviations from the Beaten Track*, p. 283)

Find and pursue something interesting that delights you especially, so you become a kind of temporary expert in some phenomenon that you heard about. It's the way to save your soul — then you can always say, "Well, at least the other guys don't know anything about this!"

– *Feynman's Tips on Physics*, p. 41

To know when you know, and when you don't know, and what it is you know and what it is you don't know, you gotta be very careful not to confuse yourself.

– Yorkshire Television interview, "Take the World from Another Point of View," 1972

You should, in science, believe logic and arguments carefully drawn and not authorities.

– Letter to Beulah E. Cox, September 1975 (*Perfectly Reasonable Deviations from the Beaten Track*, p. 290)

Study hard what interests you the most in the most undisciplined, irreverent and original manner possible.

– Letter to J. M. Szabados, November 1965 (*Perfectly Reasonable Deviations from the Beaten Track*, p. 206)

Work as hard and as much as you want on the things you like to do the best. Try to keep the other grades from going to zero if you can.

– Letter to V. A. Van Der Hyde, July 1986 (*Perfectly Reasonable Deviations from the Beaten Track*, p. 415)

It is wonderful if you can find something you love to do in your youth which is big enough to sustain your interest through all your adult life. Because, whatever it is, if you do it well enough (and you will, if you truly love it), people will pay you to do what you want to do anyway.

– Letter to Eric W. Leuliette, September 1984 (*Perfectly Reasonable Deviations from the Beaten Track*, p. 369)

For some people when you are young you only want to go as fast, as far, and as deep as you can in one subject — all the others are neglected as being relatively uninteresting. But later on when you get older you find nearly everything is really interesting if you go into it deeply enough. Because what you learned as a youth is that one thing is even more interesting as you go deeper.

– Letter to Mr. V. A. Van Der Hyde, July 1986 (*Perfectly Reasonable Deviations from the Beaten Track*, p. 283)

Work hard to find something that fascinates you. When you find it, you will know your lifework.

– Letter to student Mike Flasar, November 1966 (*Perfectly Reasonable Deviations from the Beaten Track*, p. 229)

Don't think of what "you want to be," but what you "want to do."

– Letter to V. A. Van Der Hyde, July 1986 (*Perfectly Reasonable Deviations from the Beaten Track*, p. 415)

I have just one wish for you: the good luck to be somewhere where you are free to maintain the kind of integrity I have described, and where you do not feel forced by a need to maintain your position in the organization, or financial support, or so on, to lose your integrity. May you have that freedom.

– "Cargo Cult Science," Caltech commencement address, 1974

[On social elegance of people:] It isn't true that it's very fancy and everybody believes in it, and that you're some kind of a funny guy because you don't quite get in the swing. There are a lot of people in it that are smiling, or that even are elegant and superior, tha understand what it is — that it's a show. But you don't know that at first.

– Interview with Charles Weiner, March 5, 1966 (Niels Bohr Library and Archives with the Center for the History of Physics)

Do not read so much, look about you and think of what you see there.

– Letter to Master Ashok Arora, January 1967 (*Perfectly Reasonable Deviations from the Beaten Track*, p. 230)

You cannot develop a personality with physics alone; the rest of life must be worked in.

– Letter to Mr. Alan Woodward, March 1982 (*Perfectly Reasonable Deviations from the Beaten Track*, p. 345)

A man may be digging a ditch for someone else, or because he is forced to, or is stupid such a man is "toolish" but another working even harder may not be recognized as different by the bystanders but he may be digging for treasure. So dig for treasure, and when you find it you will know what to do.

– Letter to student Mike Flasar, November 1966 (*Perfectly Reasonable Deviations from the Beaten Track*, p. 229)

Remain a thoughtful person and you will remain free. For freedom is a consistency of thought and action.

– Notes for a commencement speech

Do not remain nameless to yourself — it is too sad a way to be. Know your place in the world and evaluate yourself fairly, not in terms of the naive ideals of your own youth, nor in terms of what you erroneously imagine your teacher's ideals are.

– Letter to Koichi Mano, February 1966 (*Perfectly Reasonable Deviations from the Beaten Track*, p. 201)

I can't believe I ever said I like to hear from old friends — or maybe I did — but I hate to answer letters!

– Letter to Dr. N. H. Spector, September 1985

No problem is too small or too trivial if we can really do something about it.

– Letter to Koichi Mano, February 1966 (*Perfectly Reasonable Deviations from the Beaten Track*, p. 201)

I entered MIT in mathematics, changed to electrical engineering for a while and then settled in physics. What field of physics? Aside from deciding I liked theoretical work best, I have wandered around from stresses in molecules to quantum theory electrodynamics, theory of liquid helium, nuclear physics, turbulence in the flow of water (I didn't succeed in the last two problems so nothing is published), and recently particle physics. You do any problem that you can, regardless of field.

– Letter to student Mark Minguillon, August 1976 (*Perfectly Reasonable Deviations from the Beaten Track*, pp. 306, 309)

What do I advise? Forget it all. Don't be afraid. Do what you get the greatest pleasure from. Is to build a cloud chamber? Then g on doing things like that. Develop your talents wherever they may lead. Damn the torpedoes full speed ahead!

> – Letter to student Frederich Hipp, April 1961 (*Perfectly Reasonable Deviations from the Beaten Track*, p. 120)

Tell your son to stop trying to fill your head with science — for to fill your heart with love is enough.

> – *No Ordinary Genius*, p. 161

We've learned from experience that the truth will come out. Other experimenters will repeat your experiment and find out whether you were wrong or right. Nature's phenomena will agree or they'll disagree with your theory. And, although you may gain some temporary fame and excitement, you will not gain a good reputation as a scientist if you haven't tried to be very careful in this kind of work. And it's this type of integrity, this kind of care not to fool yourself, that is missing to a large extent in much of the research in cargo cult science.

> – "Cargo Cult Science," Caltech commencement address, 1974

It is therefore of first-rate importance that you know how to "triangulate" — that is, to know how to figure something out from what you already know. It is absolutely necessary.

> – *Feynman's Tips on Physics*, p. 39

The best thing to do is what seems best to you, not me, so that you will motivate your students by your own confidence, interest, and personality. It is not good to seek criticism from other personalities.

> – Letter to Professor P. Mitra, December 1973

If you feel that you or your friends know enough to give advice, suggest curricula, etc., that is your business. If not, then mind your own business, and go home and give your own physics students the best damn course you can.

– Letter to John M. Fowler, March 1966

We are going to suppose something: that all the energies are positive. If the energies were negative we know that we could solve all our energy problems by dumping particles into this pit of negative energy and running the world with the extra energy.

– Dirac Memorial Lectures, "The Reason for Antiparticles," 1986

Intelligence

After all, I was born not knowing and have only had a little time
to change that here and there.

- Letter to Armando Garcia, December 1965 (*Perfectly
  Reasonable Deviations from the Beaten Track*, p. 396)

Ordinary fools are all right, you can talk to them, and try to help them out. But pompous fools — guys who are fools and are covering it all over and impressing people as to how wonderful they are with all this hocus-pocus, THAT, I CANNOT STAND!

– *Surely You're Joking, Mr. Feynman!*, p. 284

There's no talent, no special ability to understand quantum mechanics, or to imagine electromagnetic fields, that comes without practice and reading and learning and study. I was born not understanding quantum mechanics — I still don't understand quantum mechanics!

– BBC, "Fun to Imagine" television series, 1983

The whole idea that the average person is unintelligent is a very dangerous idea. Even if it's true, it shouldn't be dealt with they way it's dealt with.

– "The Unscientific Age," John Danz Lecture Series, 1963

I don't know what it is exactly, but it's interesting how when you do something foolish, you protect yourself from knowing of your own foolishness.

– "The Remarkable Dr. Feynman," *Los Angeles Times Magazine*, April 20, 1986

I'm a very curious man, and I watch phenomena that happen all the time. You often wonder what it would be like to be going a little crazy. And I had this experience of going crazy, or of something wrong with my mind, and I didn't notice it. The same mind that is weakening has lost its analytical ability to watch itself. So I was simply rationalizing every failure. I didn't have the sense to realize what was perfectly obvious: A person doesn't get old in a week!

– UCSB talk, "Los Alamos from Below," February 1975

Even if you're one of the last couple of guys in the class, it doesn't mean you're not any good. You just have to compare yourself to a reasonable group, instead of to this insane collection that we've got here at Caltech.

– *Feynman's Tips on Physics*, p. 18

The first one has to do with whether a man knows what he is talking about, whether what he says has some basis or not. And my trick that I use is very easy. If you ask him intelligent questions — that is, penetrating, interested, honest, frank, direct questions on the subject, and no trick questions — then he quickly gets stuck. It is like a child asking naive questions. If you ask naive but relevant questions, then almost immediately the person doesn't know the answer, if he is an honest man.

– "This Unscientific Age" John Danz Lecture Series, 1963

We are not that much smarter than each other.

– Quoted in the *New York Times Magazine*, James Gleick, September 26, 1992

I'm an explorer, okay? I get curious about everything, and I want to investigate all kinds of stuff.

– BBC, "No Ordinary Genius," 1993

What I cannot create, I do not understand.

– Written on blackboard when he died

I've learned to draw and I read a little bit, but I'm really still a very one-sided person and I don't know a great deal. I have a limited intelligence and I use it in a particular direction.

– *What Do You Care What Other People Think?*, p. 11

No matter how carefully we select the men [only men were admitted to Caltech in 1961], no matter how patiently we make the analysis, when they get here something happens: It always turns out that approximately half of them are below average! Of course you laugh at this because it's self-evident to the rational mind, but not to the emotional mind — the emotional mind can't laugh at this. When you've lived all the time as number one or number two (or even possibly number three) in high school science, and when you know that everybody who's below average in the science courses where you came from is a complete idiot, and now you suddenly discover that you are below average — and half of you guys are — it's a terrible blow, because you imagine that it means you're as dumb as those guys used to be in high school, relatively. That's the great disadvantage of Caltech: that this psychological blow is so difficult to take.

– *Feynman's Tips on Physics*, p. 17

The Nobel Prize

I don't like honors. I appreciate it for the work that I did, and I know that there's a lot of physicists who use my work. I don't need anything else. I don't think there's any sense to anything else. I don't see that it makes any point that someone in the Swedish Academy decides that this work is "noble enough" to receive a prize. I've already got the prize: The prize is the pleasure of finding the thing out, the kick in the discovery, the observation that other people use it. Those are the real things.

– BBC, "The Pleasure of Finding Things Out," 1981

You know, I did this work in 1949. I guess they just ran out of winners and looked back over the old stuff.

– *South Shore Record*, October 28, 1965

And so, you Swedish people, with your honors, and your trumpets, and your king — forgive me. For I understand at last — such things provide entrance to the heart. Used by a wise and peaceful people, they can generate good feeling, even love, among men, even in lands far beyond your own. For that lesson, I thank you. *Tack!*

– From Les Prix Nobel en 1965 [Nobel Foundation], Stockholm, 1966

So what happened to the old theory that I fell in love with as a youth? Well, I would say it's become an old lady, that has very little attractive left in her and the young today will not have their hearts pound anymore when they look at her. But, we can say the best we can for any old woman, that she has been a very good mother and she has given birth to some very good children. And, I thank the Swedish Academy of Sciences for complimenting one of them. Thank you.

– From *Nobel Lectures, Physics 1963–1970*, Elsevier Publishing Company, Amsterdam, 1972

One of the pleasant things about winning the prize is to hear from former students.

– Letter to Loren A. Page, November 1965

I was delighted too when I heard about the Nobel Prize, thinking as you did that my bongo playing was at last recognized.

– Letter to Sandra Chester (*Perfectly Reasonable Deviations from the Beaten Track*, pp. 163–164)

There were all kinds of things, serious and humorous, telegrams and letters. And in each of them I saw happiness on the part of the people who were sending it and some real feeling of affection, which kind of overwhelmed me and made me feel real love for all these people, because they all seemed to be so good-hearted and so happy about the congratulations. I didn't realize that to have everything come at once like that, it really makes you feel good. So that was the good part of the whole thing, the letters. That was the good part.

– Interview with Charles Weiner, June 28, 1966 (Niels Bohr Library and Archives with the Center for the History of Physics)

We have not quantitatively checked this theory with the gluons — it might be wrong. We have only got a few experiments to check the W-Boson — that might be wrong. On the other hand, why does it look like it could be the same thing repeated? One: the limit of the imagination of man. When he sees a new theory and a new phenomenon, he tries to fit it with that theory, and until he's done enough experiments, he doesn't know that it doesn't work. So when he gives a lecture in 1979 in New Zealand, he thinks it works!

– "QED: New Queries," Sir Douglas Robb Lectures, University of Auckland, 1979

Some guy who made a lot of money on dynamite wants to make himself a big thing and put his name on the big prize so everybody will remember the name "Nobel," and for that I got to be annoyed. The hell with it.

> – "Nobel Prize: Another Side of the Medal," *The Los Angeles Times*, October 7, 1983

Appearances, dinner with the King, meet the King, get the Prize, tatata, all this stuff, see. And the worst of it was that I ridicule kings and things like that. I ridicule ceremony. I used to. I still do. I laugh at it. And here I have to be a party to it. It's not very consistent to laugh at it when somebody else is doing it, but when you're in it, because you're receiving a prize and so on, you're going to go right along without some kind of — You know, you used to laugh, and here you are, the Big Boy, right in the middle of it, not laughing anymore, ha ha ha!

> – Interview with Charles Weiner, June 28, 1966 (Niels Bohr Library and Archives with the Center for the History of Physics)

The real mistake that one makes when one wins the prize is to take it all too seriously for example, this speech. I worried very hard — is it appropriate to give such a speech? It don't make a goddamn bit of difference. It isn't really very serious. It doesn't make any difference what you say. After all, may I remind you that I have never in my life read the Nobel lecture of anybody? They're published, but who reads them?

> – Interview with Charles Weiner, June 28, 1966 (Niels Bohr Library and Archives with the Center for the History of Physics)

If somebody says, "We need a Nobel Prize winner to sign a letter to Russia on the Jews," I say, "I'm willing to sign a letter to Russia on the Jews but I'm not willing to be a Nobel Prize winner signing a letter to Russia on the Jews."

– Interview with Charles Weiner, February 4, 1973 (Niels Bohr Library and Archives with the Center for the History of Physics)

Now it turns out that after you win the prize, you're supposed to give a talk on what you did to win the prize. I should think that they knew the reason when they gave you the prize, but apparently they're a bit uncertain or something.

– CERN talk, December 1965

[On learning of the Nobel Prize:] The telephone rang, the guy said I'm from some broadcasting company. I was very annoyed to be awakened. That was my natural reaction. You know, you're half awake, and you're annoyed. So the guy says, "We'd like to inform you that you've won the Nobel Prize." And I'm thinking to myself — I'm still annoyed; see — it didn't register. So I said, "You could have told me that in the morning." So he says, "I thought you'd like to know." Well, I said I was asleep and put the telephone back.

– "Nobel Prize: Another Side of the Medal," *The Los Angeles Times*, October 7, 1983

I picture myself as an ordinary guy, and I hate to be discovered I hate to find that the way I picture myself isn't the way they're picturing me. They have me as some kind of Nobel Prize winner, but I still really haven't caught on that I'm any different than I was before.

    – Future for Science interview

It bothers me that everyone always chooses "Nobel prize winners" as important examples of scientists. Why do we pay such attention to the choice of the members of the Swedish Academy? That may be OK for the unknowing public, but surely a science teacher can make his own independent choices of which scientists excite his imagination and which men he would like to call to the attention of his students.

    – Letter to Mr. Stuart Zimmer, February 1982 (*Perfectly Reasonable Deviations from the Beaten Track*, p. 340)

I just get tired of being a "Nobel Prize Winner" from time to time.

    – Letter to Mr. Stuart Zimmer, February 1982 (*Perfectly Reasonable Deviations from the Beaten Track*, p. 340)

It is very hard to be pushed out of the ivory tower; the light is so strong that it hurts. What hurts worse is to think of myself in tails receiving something from the King of Sweden, while the television cameras are watching.

    – Letter to Betsy Holland Gehman, November 1965 (*Perfectly Reasonable Deviations from the Beaten Track*, p. 187)

Well isn't it exciting to hear from old school mates; that is one of the best parts of winning the Nobel Prize. All kinds of people that

I used to know and like, that I haven't heard from in a long time, come out of the woodwork.

– Letter to Wanna M. Hecker, November 1965

I used to be able to go to any old high school and answer questions at the physics clubs. But now, they don't even ask me. They're afraid. They wouldn't ask a Nobel Prize winner to talk to a physics club. And if some student finally gets up the nerve to do it, what happens is, I say OK. And when I go there, it's not just the physics club but the whole damn school is there. The principal finds out or the physics teacher finds out what the kid in the physics club has done, and they say, "Oh, he's such an important man, everybody should be interested in this guy." It's kind of out of proportion. I'm not up to it.

– "The Remarkable Dr. Feynman," *Los Angeles Times Magazine*, April 20, 1986

[On winning the Nobel Prize:] You can't escape it. A guy calls you up in the middle of the night, and my first reaction was, I won't accept. But then I realized that if I said that I'd make a bigger stink than if I took it. You're stuck. It's not fair to be stuck like that. There's no reason why your privacy and everything has to be interfered with.

– "Nobel Prize: Another Side of the Medal," *The Los Angeles Times*, October 7, 1983

[After Nobel Prize:] We had hundreds of letters, from friends all over the world, and relatives — like a relative of mine happened to be on a ship, you know, going from Spain to somewhere and, oh gee, he practically busted a gasket, and sent a big telegram. I got telephone calls from Mexico City that I can't hear because the telephone system was no good. I still tried to get back and tell that person I really liked it, and thank them for the call, but I don't know the address so I'm stuck. It was hard to hear but I finally understood who it was. All kinds of crazy stuff. Very nice letters. They were all full of — kind of happy. Everybody was excited. Each letter indicated some excitement in the house, whoever it was.

– Interview with Charles Weiner, June 28, 1966 (Niels Bohr Library and Archives with the Center for the History of Physics)

At the dancing afterwards — you see, I had to get some release from the formalities — I overdid it then. When I got informal, I just went wild, you see. So when the dancing began, we started, I danced with my wife; then I danced with somebody else, a sister of a Nobel Prize winner. I didn't get to dancing with the Princess, because I had a — you know, I wouldn't even try.

– Interview with Charles Weiner, June 28, 1966 (Niels Bohr Library and Archives with the Center for the History of Physics)

When I danced with my wife, when I danced with the daughter of a Nobel Prize winner, they were taking pictures, all the time — click, flash, flash. When I danced with this girl, which I did twice as many dances as I did with anybody else altogether — no pictures. Nothing. Not in the paper. Not a picture. Nothing. Apparently there's something wrong with this, you see, and they protect the Nobel Prize winners from their dumb idiosyncrasies. But this was my idea of relaxing, of informality. I had to do something because I had to get out from under, you know what I mean? It was fun. It was funny.

– Interview with Charles Weiner, June 28, 1966 (Niels Bohr Library and Archives with the Center for the History of Physics)

The prize was a signal to permit them to express, and me to learn about, their feelings. Each joy, though transient thrill, repeated in so many places, amounts to a considerable sum of human happiness. And, each note of affection released thus one upon another has permitted me to realize a depth of love for my friends and acquaintances, which I had never felt so poignantly before.

– From Les Prix Nobel en 1965 [Nobel Foundation], Stockholm, 1966

Hell, if I could explain it to the average person, it wouldn't have been worth the Nobel prize.

– *People*, July 22, 1985

[To a reporter after being woken up to be told about his winning the Nobel Prize:] This is a heck of an hour. I could have found out later this morning.

– *California Tech*, Caltech student newspaper, October 1965

Yeah, "give us a quote" is really what they're trying to say. And I couldn't figure a way of saying it. I gradually developed a way but it was rather too late — saying I'd worked on the interaction of radiation and matter. That sounds good and doesn't say anything.

– Interview with Charles Weiner, June 28, 1966 (Niels Bohr Library and Archives with the Center for the History of Physics)

Worldview

The first principle is that you must not fool yourself — and you are the easiest person to fool.

– "Cargo Cult Science," Caltech commencement address, 1974

Cleverness, however, is relative.

– Audio recording of Feynman Lectures on Physics, Lecture 4,
October 6, 1961

All the time you're saying to yourself, "I could do that, but I won't" which is just another way of saying that you can't.

– *Surely You're Joking, Mr. Feynman!*, p. 68

I learned very early the difference between knowing the name of something and knowing something.

– *What Do You Care What Other People Think?*, p. 14

It is necessary, I believe, to accept this idea, not only for science but also for other things; it is of great value to acknowledge ignorance. It is a fact that when we make decisions in our life, we don't necessarily know that we are making them correctly; we only think that we are doing the best we can — and that is what we should do.

– "The Relation of Science and Religion," May 1956

There is no authority who decides what is a good idea.

– "The Uncertainty of Science," John Danz Lecture Series, 1963

What Do You Care What Other People Think?

   – What Do You Care What Other People Think?

It's good to know I have such loyal, permanent rooters no matter what I do, good or bad.

   – Letter to Evie Frank, December 1965

In order to talk to each other, we have to have words, and that's all right.

   – National Science Teachers Association Fourteenth Convention lecture, "What Is Science?" April 1966

Every morning at six I have this silly habit of going out and jogging slowly (6 mph) around for 5 or 6 miles. I haven't figured out why — I don't know whether it makes me feel good or what. I always feel good, but I did before I started jogging too.

   – Letter to Mariela Johansen, January 1975

There has been much talk about the way scientists look at love and so on, and I think it isn't really quite right, that science is not a dull, hard, cold business, but as a matter of fact I believed then and I still believe, that if used right it gives you a way of looking at the world and at the meaning of things that are happening to you that gives you some control and calmness in otherwise difficult situations, and so on.

   – Interview with Charles Weiner, March 5, 1966 (Niels Bohr Library and Archives with the Center for the History of Physics)

I have a philosophy that it doesn't do any good to go and make regrets about what you did before but to try and remember how you made the decision at the time.

– Interview for Viewpoint

[On Hans Bethe:] Like most Europeans he was a very serious man. And that means that he thinks better, it's legitimate to talk about intellectual subjects at a beer party. That's all it means.

– CERN talk, December 1965

We are lucky to live in an age in which we are still making discoveries.

– *The Character of Physical Law*, p. 127

If you thought that science was certain — well, that is just an error on your part.

– *The Character of Physical Law*, p. 77

It does not do harm to a mystery to know a little about it.

– *Feynman Lectures on Physics*, vol. 1, p. 3

The brotherhood we are attempting this week is not a viewpoint to have for seven days. Nor is it simply a concept to be kept in mind for all time. Brotherhood must be a program of action, a program of real and active cooperation between the peoples of the world. And this cooperation is not simply a desirable thing.

Cooperation is a necessity for the survival of a just fraction of the vast population of the earth — you, and I, included.

– Notes from before Los Alamos

I am curious on how reliable old memories are and wonder how much we make up in our own mind when we review events. Maybe we remember saying what we would have liked to say.

– Letter to Dr. Judah Cahn, March 1983 (*Perfectly Reasonable Deviations from the Beaten Track*, p. 362)

And there's something that struck me, it's very curious: I suspect that what goes on in every man's head might be very, very different.

– BBC, "Fun to Imagine" television series, 1983

I share sorrow in the loss of a man's life, which added significance and interest to all human life.

– In a telegram to Dr. Aage Bohr, regarding the death of Dr. Neils Bohr

If in some cataclysm, all of the scientific knowledge is to be destroyed and only one sentence is to be passed on to the next generations of creatures, what would be the best thing, the thing that contains the most information in the least number of words? believe it is the atomic hypothesis, or the atomic fact, or whatever you want to call it, that all things are made out of atoms, little particles that move around, are in perpetual motion, attract each other when they are some distance apart, but repel being squeezed into one another. In that one sentence, you'll see that there is an enormous amount of information about the world if just a little imagination and thinking is applied.

– Audio recording of Feynman Lectures on Physics, Lecture 1, September 26, 1961

The worthwhile problems are the ones you can really solve or help solve, the ones you can really contribute something to. A problem is grand in science if it lies before us unsolved and we see some way for us to make a little headway into it. I would advise you to take even simpler, or as you say, humbler, problems until you find some you can really solve easily, no matter how trivial. You will get the pleasure of success, and of helping your fellow man, even if it is only to answer a question in the mind of a colleague less able than you.

– Letter to Koichi Mano, February 1966 (*Perfectly Reasonable Deviations from the Beaten Track*, p. 198)

The world is a dynamic mess of jiggling things if you look at it right.

– BBC, "Fun to Imagine!" television series, 1983

What I am trying to do is bring birth to clarity, which is really a half-assedly thought-out pictorial semi vision thing.

– Quoted in James Gleick, *Genius: The Life and Science of Richard Feynman*, 1992, p. 244

The grand ages of different civilizations are characterized by people's enormous confidence in success, their belief that they have some new thing that is different, and their certainty that they are developing it by themselves.

– MIT centennial, "Talk of Our Times," December 1961

It is quite certain that many things are inherited but it is evil and dangerous to maintain, in these days of little knowledge of these matters, that there is a true Jewish race or specific Jewish hereditary character.

– Letter to Tina Levitan, February 1967 (*Perfectly Reasonable Deviations from the Beaten Track*, p. 235)

To select, for approbation, the peculiar elements that come from some supposedly Jewish heredity is to open the door to all kinds of nonsense on racial theory. Such theoretical views were used by Hitler. Surely you cannot maintain on the one hand that certain valuable elements can be inherited from the "Jewish people," and deny that other elements which other people may find annoying or worse are not inherited by these same "people."

– Letter to Tina Levitan, February 1967 (*Perfectly Reasonable Deviations from the Beaten Track*, p. 235)

The error of anti-Semitism is not that the Jews are not really bad after all, but that evil, stupidity and grossness is not a monopoly of the Jewish people but a universal characteristic of mankind in general.

– Letter to Tina Levitan, February 1967 (*Perfectly Reasonable Deviations from the Beaten Track*, p. 235)

[On his mental health after the bomb:] I would see people building a bridge and I would say, "They don't understand." I really believed that it was senseless to make anything because it would all be destroyed very soon anyway, but they didn't understand that. And I had this very strange view of any construction that I would see. I always thought how foolish they are to try to make something. So I was really in a kind of depressive condition.

  – BBC, "The Pleasure of Finding Things Out," 1981

The error of pro-Semitism is not that the Jewish people or the Jewish heritage is not really good, but rather the error is that intelligence, good will, and kindness is not, thank God, a monopoly of the Jewish people but a universal characteristic of mankind in general.

  – Letter to Tina Levitan, February 1967 (*Perfectly Reasonable Deviations from the Beaten Track*, p. 235)

The way my two children responded to my stories was very different, but I don't know that it was because one was a boy and the other a girl. I think people are very different and that if I had two sons they would respond differently too — maybe.

  – Letter to Dorothy Weeks, February 1983 (*Perfectly Reasonable Deviations from the Beaten Track*, p. 358)

The Future

Why repeat all this? Because there are new generations born every day. Because there are very great ideas developed in the history of man, and these ideas do not last unless they are passed purposely and clearly from generation to generation.

– "The Uncertainty of Science," John Danz Lecture Series, 1963

When we get stuck in a certain place, it's a place where History will not repeat herself. And that makes it even more exciting, because whatever we're going to look at — the method and the trick and the way it's going to look — is going to be very different than anything we've seen before, because we've used all the methods from before.

– Yorkshire Television interview, "Take the World from Another Point of View," 1972

If we thought the past was a long time, the future seems incomprehensibly longer.

– From notes for "About Time" program, 1957

We can always say to other people: "That was very clever of you to have explained why the world just has to be the way we have found it to be so far. But what is it going to look like tomorrow?" Our philosophy's vigor comes from the fact that we are still struggling.

– MIT centennial, "Talk of Our Times," December 1961

Fundamental physics has a finite lifetime. It has a while to go. At the present moment, it is going with terrific excitement, and I do not want to retire. But I'm taking advantage of the fact that I live at the right age.

– MIT centennial, "Talk of Our Times," December 1961

We like improved production, but we have problems with automation. We are happy with the development of medicine, and then we worry about the number of births and the fact that no one dies from the diseases we have eliminated. Or else, with the same knowledge of bacteria, we have hidden laboratories in which men are working as hard as they can to develop diseases for which no one else will be able to find a cure. We are happy with the development of air transportation and are impressed by the great airplanes, but we are aware also of the severe horrors of air war. We are pleased with the ability to communicate between nations, and then we worry about the fact that we can be snooped upon so easily. We are excited by the fact that space can now be entered; well, we'll undoubtedly have difficulty there too. The most famous of all of these is of course the development of nuclear energy and its obvious problems.

– "The Uncertainty of Science," John Danz Lecture Series, 1963

We are here only at the very beginning of time for the human race. There are thousands of years in the past, and there is an unknown amount of time in the future. There are all kinds of opportunities, and there are all kinds of dangers.

– "The Uncertainty of Values," John Danz Lecture Series, 1963

From a long view of the history of mankind, seen from, say, ten thousand years from now, there can be little doubt that the most significant event of the 19th century will be judged as Maxwell's discovery of the laws of electrodynamics. The American Civil War will pale into provincial insignificance in comparison with this important scientific event of the same decade.

– *Feynman Lectures on Physics*, pp. 1–11

I am sorry to have to answer your question (as to whether I consider nuclear energy a curse or a salvation of mankind) that I don't really know. I look to the future neither with hope nor fear but with uncertainty as to what will be.

> – Letter to Dr. David A. Marcus, February 1975 (*Perfectly Reasonable Deviations from the Beaten Track*, p. 279)

I hope for freedom for future generations, freedom to doubt, to develop, to continue the adventure of finding out new ways of doing things, of solving problems.

> – "The Uncertainty of Values," John Danz Lecture Series, 1963

We are only at the beginning of the development of the human race; of the development of the human mind, of intelligent life, we have years and years in the future. It is our responsibility not to give the answer today as to what it is all about, to drive everybody down in that direction and to say, "This is a solution to it all." Because we will be chained then to the limits of our present imagination.

> – Galileo Symposium, "What Is and What Should Be the Role of Scientific Culture in Modern Society," September 1964

We are at the very beginning of time for the human race. It is not unreasonable that we grapple with problems. But there are tens of thousands of years in the future. Our responsibility is to do what we can, learn what we can, improve the solutions, and pass them on.

> – "The Value of Science," December 1955

I have often made the hypothesis that ultimately physics will not require a mathematical statement, that in the end the machinery will be revealed, and the laws will turn out to be simple, like the checker board with all its apparent complexities.

— *The Character of Physical Law*, pp. 57–58

People in the past, in the nightmare of their times, had dreams for the future. And now that the future has materialized, we see that in many ways the dreams have been surpassed, but in still more ways there are many of our dreams of today which are very much the dreams of people of the past.

— Galileo Symposium, "What Is and What Should Be the Role of Scientific Culture in Modern Society," September 1964

So there's a great deal that goes over from the one science to the other, and the most important thing that goes over is the character of the science, the critical character. It's very, very much the same. The sources of dirt and the sources of error are physically different. But you can still get an idea: does it make sense or doesn't it make sense?

— Interview with Charles Weiner, June 28, 1966 (Niels Bohr Library and Archives with the Center for the History of Physics)

Future always uncertain. Is there one?

— Notes

There are a few young men hoping to be theoretical physicis who will be inspired to take the attitude: "Those guys don't kno what the hell they've been talking about all these years, they couldn't even solve the most simple problem. I'll show 'em how to do it." That's good, that might happen.

– Caltech lecture on particles, 1973

I had a calculus book once that said, "What one fool can do, another can." What we've been able to work out about nature may look abstract and threatening to someone who hasn't studied it, but it was fools who did it, and in the next generation, all the fools will understand it.

– *Omni* interview, February 1979

Each generation that discovers something from its experience must pass that on, but it must pass that on with a delicate balance of respect and disrespect, so that the race that is now aware of the disease to which it is liable does not inflict its errors too rigidly on its youth, but it does pass on the accumulated wisdom, plus the wisdom that it may not be wisdom.

– National Science Teachers Association Fourteenth Convention lecture, "What Is Science?" April 1966

If one were walking through a building to get from one side to the other and had not yet reached the door, one might argue, "Look, we have been walking through this building, we have not reached the door; therefore, there is no door at the other end." It seems to me that we are walking through a building, but we do not know whether it is an infinite building or a finite building, so there is still a possibility of a final solution. One thing that would happen, I think, if a final solution were found would be deterioration in the philosophy of science.

– MIT centennial, "Talk of Our Times," December 1961

My son is like that, too, although he's much wider in his interests than I was at his age. He's interested in magic, in computer programming, in the history of the early church, in topology — oh, he's going to have a terrible time, there are so many interesting things.

– *Omni* interview, February 1979

Honoring Richard Feynman

He was the most original mind of his generation.

> – Freeman Dyson of the Institute for Advanced Study in Princeton,
> *New York Times*, February 17, 1988

He's the most creative theoretical physicist of his time and a true genius. He has touched with his unique creativity just about every field of physics.

– Sidney D. Drell, former president of the American Physical Society, *New York Times*, February 17, 1988

In science, as well as in other fields of human endeavor there are two kinds of geniuses: the "ordinary" and the "magicians." An ordinary genius is a fellow that you and I would be just as good as, if we were only many times better. There is no mystery as to how his mind works. Once we understand what he has done, we feel certain that we, too, could have done it. It is different with the magicians . . . . Even after we understand what they have done it is completely dark . . . . Richard Feynman is a magician of the highest caliber.

– Marc Kac, *Enigmas of Chance*, p. xxv

C.P. Snow described Feynman "as though Groucho Marx was suddenly standing in for a great scientist."

– "The Cult of Richard Feynman," *Los Angeles Times Magazine*, December 2, 2001

would drop everything to hear him lecture on the municipal drainage system.

– David Mermin of Cornell, cited in *Lectures on Computation*, ed. Tony Hey

Dick made a conscious effort to look at problems from a different vantage point. It was deliberate.

> – Thomas A. Tombrello, Caltech Oral History Archives, interview by Heidi Aspaturian, December 2010

I'm thinking how to measure how smart Feynman was, because it wasn't any standard sort of smart. It was this way of viewing the world obliquely, and he tried to get there deliberately. I think he worked really hard at that and succeeded in marvelous ways.

> – Thomas A. Tombrello, Caltech Oral History Archives, interview by Heidi Aspaturian, December 2010

Well, Dick Feynman reinvented the wheel, but it turned out to be a much better wheel.

> – Valentine L. Telegdi, Caltech Oral History Archives, interview by Sara Lippincott, March 2002

Feynman was not a theorist's theorist, but a physicist's physicist and a teacher's teacher.

> – Valentine L. Telegdi, *Physics Today*, February 1989

He is a second Dirac, only this time human.

> – Eugene Wigner, quoted by Robert Oppenheimer in a letter to Professor Raymond Birge, University of California, Berkeley, 1943

An honest man, the outstanding intuitionist of our age, and a prime example of what may lie in store for anyone who dares to follow the beat of a different drum.

– Julian Schwinger, *Physics Today*, February 1989

I knew Feynman a bit, and I will say that Feynman could do in a day what 100 John Rigdens could never do. Never do.

– John Rigden, interview with Dr. Dudley Herschbach, American Institute of Physics, 2003

He had a great reputation. He was already heralded as this very clever fellow from Princeton who knew everything. And he did know everything, you know. He did solve some problems for us, just like that.

– Philip Morrison, interview with Charles Weiner, American Institute of Physics, 1967

You can talk to Feynman and his answers are precise and of the type that an experimental physicist can understand, or at least thinks he understands.

– Carl Anderson, Caltech Oral History Archives, interview by Harriett Lyle, 1979

When Dick was faced with a problem in mathematics, he was tremendously intuitive. He would find ways to solve the problem, or to prove something that he'd conjectured was true. These ways would be quite original and would usually be considered completely unorthodox by the mathematical fraternity. But they worked. He understood mathematics well enough to invent new mathematics that was intrinsically correct. He didn't make mistakes at it; it's just that he developed novel ways to do things that fitted in with his experience, and achieved results that sometimes took other people quite a while to understand how he'd achieved them.

– Charles A. Barnes, interview with Heidi Aspaturian, Caltech
  Oral History Archives, July–August 1987

Before he went off to Sweden to receive the Nobel Prize, he gave an absolutely wonderful lecture on the campus to the local people in the little theatre — Culbertson — that used to exist here on campus. It was a beautiful little building, and it held two or three hundred people, I guess. We were just fascinated while Dick made a very clear and extremely modest, typical Dick Feynman exposition of how he had got to the position that finally ended up with him being invited to Sweden to get the Nobel Prize. He explained that the way he got to the point of being able to formulate his rules for quantum electrodynamics was by doing every difficult quantum electrodynamics problem that people would bring to him.

– Charles A. Barnes, interview with Heidi Aspaturian, Caltech
  Oral History Archives, July–August 1987

But whenever Dick went somewhere for discussions of difficulties with then standard theory, he would invite people to dig out problems that they couldn't solve or could only solve with incredible difficulty. The ones that were particularly useful to him were those that other people had solved, but only with enormous difficulty. By looking at these problems and learning how to solve them by his own methods, Dick evolved his own set of rules.

– Charles A. Barnes, interview with Heidi Aspaturian, Caltech Oral History Archives, July–August 1987

I remember Richard Feynman saying, "I could never figure out what all these other guys were doing, therefore I did it my own way." And, you know — to paraphrase Frost — that made all the difference with Dick.

– Thomas A. Tombrello, interview with Heidi Aspaturian, Caltech Oral History Archives, December 2010

Truly one of the most original people at Caltech.

– Thomas A. Tombrello, interview with Heidi Aspaturian, Caltech Oral History Archives, December 2010

With Dick, there was the Feynman effect. It's like the Chinese restaurant effect — ten minutes after dinner you're hungry again. With Dick, the lecture was so clear that you quit taking notes. And then five minutes after the lecture, you couldn't reproduce the lecture! I remember when Matt Sands and Leighton, people like that, were taking notes for the Feynman lectures in freshman physics. They often realized at the end of a talk that they couldn't reproduce it. They had photographs of the board. They had recorded what Feynman said. Still, there was something elusive about it. I'm not saying it was wrong or incomplete. It was subtle.

And you didn't realize the subtlety, because it was so smooth, it was so beautifully done. It was a piece of artwork. But you had to constantly be aware of the fact that because Dick made it seem so simple, you were missing key things. The Feynman effect. It was very interesting.

– Thomas A. Tombrello, interview with Heidi Aspaturian, Caltech Oral History Archives, December 2010

I had a very strange interaction with Feynman. We were discussing something, and he said to me, "If I don't absolutely know something about it, I don't say I know something about it." And this was meant in a very friendly way. You don't get offended by that.

– Samuel Epstein, interview with Carol Bugé, Caltech Oral History Archives, December 1985–January 1986

He made the person he was talking to feel charming and witty and suddenly feel that you could do high-level physics. And he was a wonderful listener.

– Jenijoy La Belle, interview with Heidi Aspaturian, Caltech History Archives, February–May 2008, April 2009

The students regarded him as their patron saint at Caltech, [and the reputation] is quite deserved.

– Steven C. Frautschi, interview with Shirley K. Cohen, Caltech History Archives, June 2003

I remember Feynman's story. They called him up and said, "Yo won the Einstein Prize." He said, "Well, what's that?" And the

said, "Well, you get $15,000. Don't you have anything to say?"
He said, "Hot dog!"

> – Seymour Benzer, interview with Heidi Aspaturian, Caltech
> Oral History Archives, September 1990–February 1991

I didn't know Dick well at all. I knew him well enough to call him
Dick, but that was about all; we hadn't had a lot of interaction. I
said, "Dick, what's special about the center of the galaxy? Why
should we see something like this? Is there something special
about that?" He was standing there and looking down at this
[chart on the floor]. He said, "That's where God lives."

> – James A. Westphal, interview with Shirley K. Cohen, Caltech
> Oral History Archives, July 1998

He had an enormous impact on me, not just intellectually but in kind
of seeking the truth. All these things underneath that drove him —
not just how smart he was — had a really big influence on me.

> – Barry C. Barish, interview with Shirley K. Cohen, Caltech Oral
> History Archives, July 1998

Feynman is a much more exuberant person and much more of an
extrovert, really. Life is just happiness for him, even if he's had all
these medical troubles. He just is a fundamentally happy person.

> – Hans A. Bethe, interview with Judith R. Goodstein, Caltech
> Oral History Archives, February 1982

The thing that always impressed me about Dick was that you
could ask him a question, and if it wasn't a very good question,

he would take it and turn it around and answer maybe another question that was a good question. You'd just learn an enormous amount from him.

– Alvin V. Tollenstrup, interview with David A. Valone, Caltech Oral History Archives, December 1994

And these new areas; when I was an undergraduate, nobody did quantum mechanics, except super-advanced PhD students. First they had to learn all sorts of fancy Hamiltonian mechanics and all sorts of stuff, it was thought then. You had to go through all that monkey business before you could hope to even begin quantum mechanics . . . . Richard Feynman, among other people, showed that you don't have to go through that other stuff. You just start talking about quantum mechanics and the kids lap it up.

– David S. Wood, interview with Shirley K. Cohen, Caltech Oral History Archives, May 1994

The faculty and the students are in many ways very much alike here. This I gradually began to learn in those postwar days, when everybody was new here or was restarting again after the war. The students of course, in a sense, copy the faculty, because the faculty are role models for them. Dick Feynman is the great example. They all love him, and rightly so.

– Rodman W. Paul, interview with Carol Bugé, Caltech Oral History Archives, February 1982

A lecture by Dr. Feynman is a rare treat indeed. For humor and drama, suspense and interest it often rivals Broadway stage plays. And above all, it crackles with clarity. If physics is the underlying "melody" of science, then Dr. Feynman is its most lucid troubadour.

– Irving Bengelsdorf, *Los Angeles Times* science editor, 1967

When people get Nobel Prizes, that's probably one of the chief problems that the division chairman has, trying to hold onto the best people on the staff, and trying to handle these offers that they are continually getting. And if they get a Nobel Prize, that doesn't help solve that problem. Except, I do want to say that in Feynman's case, he said he had made up his mind; he likes Caltech and he wants to stay here, no matter what offers he gets — and you can be sure he's gotten them from about every place in the world.

I heard that when he answered the phone, he said, "Are you going to give me an offer?" If the person said, "No, I'm not at all interested in that," then he'd go ahead and talk to him; otherwise he would say, "The answer's no," and hang up. Maybe that's a joke, but I heard that about Dick Feynman. He's a very loyal Caltech professor.

– Carl Anderson, interview with Harriett Lyle, Caltech Oral History Archives, January 1979

Everything in retrospect, all the physics that I use today, and it seems like 90% of it I must have learned from Feynman and I had never seen anyone work so quickly and so I had never come across a physicist like that. Certainly nobody at Princeton or at Oxford was like this. And he was pretty fast and terrible when he was young. He is not much less fast or terrible now. And, you know when an idea popped into his head it would take literally no more than 5 or 10 minutes to work this kind of thing out.

– Robert Hellwarth, interview with Joan Bromberg, May 1985, American Institute of Physics

One of the great things about Feynman was the fun it was in taking up a problem and discussing it, kicking things around, laughing. I can recall his delight in one of the tricks played when he was a student at MIT, students getting together and hoisting a car and leaving it up on the roof of one of the buildings for the administration to deal with. [laughs] And the same delight he got in opening safes. I think I've told the story about the two safes at General Electric in Schenectady that Feynman opened while the security man was looking on . . . . I think that sense of fun that Feynman had — he explained afterward about how so many people use numbers like e and pi and their license number and the telephone number. Those were the numbers that he first tried to use with the greatest chance of solving the safe opening problem.

– John Wheeler, interview with Kenneth W. Ford, American Institute of Physics, March 1994

The people I know admired his breadth of interest, his curiosity and his love of life. Most of us tend to have much narrower lives than he, and many of us regret at times that we don't have the breadth of experience and relationships he did.

– Kip Thorne, "The Cult of Richard Feynman," *Los Angeles Times Magazine* December 2, 2001, p. 16

It was going along in an orderly way, but it was clear that the outspoken people were against it. Then Feynman got up, finally, and said, "I've been thinking about this and I think it's a terrible mistake. It's not Caltech. We would be giving the wrong message. And I think we should immediately squash it. It doesn't matter whether we've made mistaken commitments. We should just not do it." And that was the vote, essentially — that was the resolution. What Feynman said carried the day. It was pretty persuasively said.

– Fred Anson, reminiscing on the vote about whether to establish an army research center at Caltech, interview with Shirley K. Cohen, Caltech Oral History Archives, February 1997

He was quite a character with a sense of humor that just wouldn't quit.

– Childhood friend Joseph Heller, interview with Shelley Erwin, Caltech Oral History Archives, May 2010

I recall that Dick used to ask his class a question. He'd say, "OK. You have a lawn sprinkler, and you turn on the water, and the sprinkler rotates while shooting out oblique jets of water. Now, suppose you put this same sprinkler in a swimming pool and start to suck the water out through it. Would the sprinkler still be rotating?" It would make a great difference — like day and night. Things like that made the category of separated flows very interesting . . . I think it was a graduate course, but I cannot remember exactly which class. I was there when he asked us this question, and I thought, "Ah, that's a great teacher." A lawn sprinkler — everybody sees those, daily. But he just changed the direction of the water flow in the hose and we had to say what would happen.

– Theodore Y. Wu, interview with Shirley K. Cohen, Caltech Oral History Archives, February–March 2002

But the one person here who had a tremendous influence on me and whom I considered unique was Feynman. He had an enormous impact on me, not just intellectually but in kind of seeking the truth. All these things underneath that drove him — not just how smart he was — had a really big influence on me.

– Barry C. Barish, interview with Shirley K. Cohen, Caltech Oral History Archives, May/July 1998

I had Dick Feynman for mathematical physics. I used to go around to all his seminars when I was an undergrad, even. I couldn't understand any of the mathematics at all, but every once in a while he'd stop and he'd say, "What this really means is . . .," and I could understand that.

– Carver Mead, interview with Shirley K. Cohen, Caltech Oral History Archives, July 1996

He liked teaching, but he didn't like to supervise. The way he told it to me once was, that if he could formulate a problem sufficiently straightforwardly for a graduate student to do for a thesis, he coul do it himself in one evening. If he could get a problem that clearly set, he couldn't refrain from doing it.

> – Robert F. Christy, interview with Sara Lippincott, Caltech Oral
> History Archives, June 1994

I think during my time on the faculty at Caltech, the interesting thing to me there was the great movement forward in the pedagogy. In teaching those core courses, so-called. In terms of physics, I think it was largely due to Feynman. It used to be thought that a student, before he could have a hope of even starting to learn anything about quantum mechanics, had to go all through a great long rigmarole of classical mechanics at very sophisticated levels and so on. And Feynman showed that that's not true. You don't have to do that. . . . The material they're given and what they learn here now is far, far advanced, compared to when I was an undergraduate. You know, it's like the difference between the Model T Ford and the latest car, or the Wright brothers' airplane and a 747.

> – David S. Wood, interview with Shirley K. Cohen, Caltech Oral
> History Archives, June 1994

Feynman invented a whole new way of doing quantum mechanics, and his diagrams didn't spring from any contact with the mathematicians; in fact, Dick has generated things. And he sometimes made remarks that mathematicians don't really help very much. He got very angry about the way mathematics was taught in the California schools. He was on a committee or something, for the governor or somebody. So you see, the whole thing turned around.

> – William A. Fowler, John Greenberg, and Carol Bugé, Caltech
> Oral History Archives, May 1983–May 1984

You can talk to Feynman and his answers are precise and of the type that an experimental physicist can understand, or at least thinks he understands.

– Carl Anderson, interview with Harriett Lyle, Caltech Oral History Archives, January 1979

I discussed this matter with Feynman once, who is perhaps that physicist in the United States who most understands this side of physics. He has this kind of attitude to see whether the thing is right. But we discussed this and he said, "Well, among the younger generation nowadays there are very few people who would dare to publish a thing which contained contradictions." Practically nobody would, because he would say, "Then I will very soon be criticized by the other fellows who would say, 'There is your contradiction, you must be wrong.'" But then to say, "Well, I know that I must be wrong; certainly there is a contradiction, but damn it, I can see that it's right." Now, of course, you can again say that that is a very funny attitude. How can you know it? You cannot prove it; it contains contradictions.

– Werner Heisenberg, interview with Thomas S. Kuhn, American Institute of Physics, February 1963

[On Feynman humor:] I have no quarrel with this aspect of Dick Feynman's personality; like most other people, I found it delightful. But it leaves out so much: not only his scientific genius, but his deep love of nature, his passion for teaching, and above all his extraordinary standards of personal integrity, no *always* present at the highest levels of creativity. . . . Whatever else Dick Feynman may have joked about, his love for physics approached reverence.

– Laurie M. Brown, *Physics Today*, February 1989

[On his personality:] I had a glimpse of it on British television years before when I had no idea who Feynman was other than an American professor who might have been invented by Arthur Miller to do monologues on such matter — and, come to that, anti-matter — in which I had no special interest, no background, and no insight, but which was made mysteriously captivating. Later on, that personality came through intoxicatingly in his taped reminiscences published as *Surely You're Joking, Mr. Feynman!*; the physicist not of the faculty club but of the saloon.

– Tom Stoppard, "Stage View," *New York Times*, November 27, 1994

I recall the time I first saw Richard, for only a few hours just forty-five years ago. He was on his way to New Mexico, I expect, passing through the Chicago Laboratory of the Manhattan district. A half-dozen theorists gathered to meet him; his reputation had preceded him from Princeton. One or two people showed him difficult integrals they hoped he could solve, a little like asking the visiting strong man to loosen some rusted gate. He performed as hoped, but it was not the kind of light in which his wonderful zest for the rhythms and the puzzles of the world could be made out. But I saw that clearly later, at Los Alamos and around our shared old office in Cornell, and I came to love and admire his amazingly original, generous, honest, and playful mind, and a spirit that seemed like his gestures to dance through life.

– Philip Morrison, personal letter of condolence, February 1988

Dick Feynman's contributions to physics have made an extraordinary impact on our work: his development of quantum electrodynamics, his invention of the path-integral formulation of quantum field theory, his essential contributions to the theory of the weak interactions, his invention of the parton model for deep inelastic electron-proton scattering, and his extraordinary insights into the nature of high energy collisions which led to the unraveling of the quark and gluon structure of matter. His work touched virtually every area of physics.

Through his teaching and remarkable personality, Feynman taught us that physics is not only profound but also intuitive and comprehensible. We are also proud of his service to his country and the courage he showed on the *Challenger* investigation. Few scientists have made more impact on science and society. We treasure his memory.

– The SLAC (Stanford Linear Accelerator Center) Theoretical Physics Group, personal letter of condolence, February 1988

In the score of years that I knew him, as weekly lecturer at the Labs every Wednesday, as critic, consultant, and very human being, he was a great inspiration, a great mind, and a great spirit. Often he would show the greatest patience and kindness in transforming an innocent but ill-informed or stupid question into a rich and brilliant one, saving the questioner from embarrassment. His great humor matched his great intellect. He will live forever in the minds of this and all future generations of physicists and of all educated people.

– Bernard Soffer, Hughes Aircraft Company Research Laboratories, personal letter of condolence, February 1988

Dick had such a dynamic and colorful personality. He had great enthusiasm for learning. He was known as one of the world's most brilliant physicists and original thinkers, yet he was concerned to make science understandable and fascinating to others, thereby making him one of Caltech's most outstanding and favorite teachers. He was very creative in explaining the most esoteric concepts so that ordinary people could understand. Feynman stories are legend here, always told with affection and admiration.

– Sunney I. Chan, Chairman of the Faculty, California Institute
   of Technology, personal letter of condolence, February 1988

Dick was the best and favorite of several "uncles" who enriched my childhood. During his time at Cornell, he was a frequent and always welcome visitor at our house, one who could be counted on to take time out from conversations with my parents and other adults to lavish attention on the children. He was at once a great player of games with us and a teacher even then who opened our eyes to the world around us.

– Henry Bethe, personal letter of condolence February 1988

Acknowledgments

There are many people to thank for their assistance.

First, my two researchers were incredible — Anisha Cook and Janna Wennberg. I was so very lucky to have your help on this project. I find it impossible to imagine I could have done this without the two of you and your hours of hard work. Thank you for your efforts.

I can always count on an honest assessment of my work from my friend Gregory Feldmeth, assistant head and long-time history teacher at Polytechnic School. I really appreciate his tireless efforts helping me with this project. He was instrumental in critically reviewing the sorting of quotes into categories, assisting with the painstaking task of removing duplicate quotes, and helping me create a timeline of my father's life. I was grateful to have his counsel and company. Polytechnic Upper School English teacher Grace Hamilton was incredibly generous with her time and expertise while assisting with the preface. Another friend, Richard White, Polytechnic School physics and computer science teacher extraordinaire, recommended the unstoppable team of Janna and Anisha, and was giving with his advice and technical ingenuity. I can understand why these three teachers are beloved by students.

Leslie Carmell, director of communications at Polytechnic School, is a dream to work with. She has been enormously supportive to me, and I am so grateful for her eagle eye and always-insightful suggestions. My colleagues are exceptional — Jennifer

Godwin Minto, Barbara Bohr, John Yen, as well as the host of talented teachers and staff at Polytechnic School. I am fortunate to work there.

Melanie Jackson, I am glad I have a wonderfully intelligent and kind literary agent such as you to advocate for me.

Thank you to my brother, Carl Feynman, for his confidence in me to complete this project.

Ralph Leighton was very accommodating and emailed audio files of *The Feynman Lectures on Physics* as he digitized them and was very, very helpful with his suggestions and advice. Christopher Sykes was impressive with his ideas and knowledge about sources of various quotes. Adam Cochran from Caltech was instrumental with communication with Caltech — everything from getting permissions to connecting me with people who were helpful, thorough, and efficient. Shelley Erwin and Loma Karklins from the Caltech archives were exactly that. Thank you for your knowledge and assistance. Tony Hey was kind enough to dig through his personal archives and scan a transcript from a panel discussion that was difficult to find. Ann Rho, senior director of development at Caltech, was eager to help, and was a good person with whom to brainstorm ideas. Alan Alda, thank you for your friendship and support. I am grateful to Kip Thorne for his guidance, not the least of which was putting me in touch with Brian Cox!

Brian Cox has a communication style reminiscent of my father's, and I am delighted he agreed to write a foreword for this book.

Yo-Yo Ma is a long-time family friend whom I'll describe as kindness personified. My father and I spent many hilarious evenings backstage and at dinner after attending Yo-Yo's concert (the tradition continues — I now go with my children), and Yo-Yo dedicated a series of concerts to my father shortly after my father died. I'm not sure why I thought it was a good idea to burden my friend who is a professional cellist to write something for this book, but I'm impressed that he didn't back down from the challenge.

The last few years have not been easy, and I have become increasingly grateful for — and at times, dependent on — my fantastic friends. It seems like a long list, and I'm only sad that I can't thank *all* the people who have meant so much to me over the last few years. I am blessed to be part of a wonderful community. I want to thank Megan Foker — I don't know where I would be without her. She helped me through a challenging period in my life and did it with such humor, grace, and equanimity that it was easier to remain balanced. Everyone needs a friend like Megan. I appreciate Rick Foker's help with their family so Megan could spend time with me. I'm proud that my community includes: Heather and Tom Unterseher, Cheryl Wold and Paul Wennberg, Electra and Peter Lang, Jane Kaczmarek, Stacy and Michael Berger, Dyanne di Rosario-Halsted and Chris Halsted, Tim Hartley and Jason Lyon, Tiffany and Marc Harris, Mario Miralles and Brenda Bork, Ralph Leighton and Phoebe Kwan, Kevin and Kristen Tyson, Scott Lee and Karen Wong, Francisco Miralles, Susan Blaisdell, Dorothy Rubin, Carl and Paula Feynman, Charles Hirshberg and Alison Adler, and Joan Feynman. You are my support team. Your friendship, love, and support have buoyed my spirits over the years. John Kurlowski, you came along at exactly the right moment and knew just how to help and support me.

Finally, I'd like to thank my children, Ava and Marco Miralles, for their fantastic ideas, burgeoning independence, and general helpful attitudes about the project. I love you.

Photo Credits

Page 241: Courtesy Michelle Feynman and Carl Feynman
Page 249: Courtesy Michelle Feynman and Carl Feynman
Page 261: Courtesy California Institute of Technology
Page 271: Courtesy California Institute of Technology
Page 281: Courtesy Michelle Feynman and Carl Feynman
Page 293: Courtesy California Institute of Technology
Page 317: Courtesy California Institute of Technology
Page 327: Courtesy Michelle Feynman
Page 333: Hughes Aircraft Company
Page 345: Courtesy Michelle Feynman and Carl Feynman
Page 355: Courtesy Michelle Feynman and Carl Feynman
Page 363: Courtesy Michelle Feynman and Carl Feynman

Sources

American Physical Society Annual Meeting, 1950

Audio recording of lecture on relativity, Douglas Advanced Research Laboratory, 1967

BBC, "Fun to Imagine" television series, 1983

BBC, "Horizon: The Hunting of the Quark," May 1974

BBC, "The Pleasure of Finding Things Out," 1981

BBC, "Strangeness Minus Three," 1964

BBC interview, "A Novel Force in Nature"

BBC interview, "Scientifically Speaking," April 1976

California Tech, Caltech student newspaper, October 1965

Caltech lecture on particles, 1973

"Cargo Cult Science," Caltech commencement address, 1974

CERN talk, December 1965

The Character of Physical Law, MIT Press, Richard P. Feynman, 1965

"The Computing Machines in the Future," Nishina Memorial Lecture, August 1985

"The Cult of Richard Feynman," Los Angeles Times Magazine, December 2, 2001

Dirac Memorial Lectures, "The Reason for Antiparticles," 1986

Esalen lecture, "Computers from the Inside Out," Esalen Institute, produced by Faustin Bray/ Sound Photosynthesis Mind@Large Catalog

Esalen lecture, "Quantum Mechanical View of Reality (Part 1)," Esalen Institute, produced by Faustin Bray/ Sound Photosynthesis Mind@Large Catalog, Z440-07, 1984

Esalen lecture, "Quantum Mechanical View of Reality (Part 2)," Esalen Institute, produced by Faustin Bray/ Sound Photosynthesis Mind@Large Catalog, Z440-07, 1984

Esalen workshop, "Tiny Machines," Esalen Institute, produced by Faustin Bray/ Sound Photosynthesis Mind@Large Catalog

Eugene Wigner, quoted by Robert Oppenheimer in a letter to Professor Raymond Birge on November 4, 1943, University of California, Berkeley

"Feynman: Frustrated by the Slow Pace of Probe," *Pasadena Star-News*, January 29, 1989

Feynman Lectures on Computation, Perseus Publishing, Richard P. Feynman, edited by Tony Hey and Robin W. Allen, 1996

Feynman Lectures on Gravitation, Addison-Wesley Publishing, Richard P. Feynman, edited by Fernando B. Morinigo and William G. Wagner, 1995

Feynman Lectures on Physics, Addison Wesley, Richard P. Feynman, edited by Robert B. Leighton and Matthew Sands, 1963

"The Feynman Legend," *The Los Angeles Times*, February 17, 1988

"Feynman Takes NASA to Task," *Pasadena Star-News*, June 11 1986

Feynman's Tips on Physics, Basic Books, Richard P. Feynman, edited by Michael A. Gottlieb and Ralph Leighton, 2013

Future for Science interview, R. P. Feynman Papers, California Institute of Technology Archives

Galileo Symposium, "What Is and What Should Be the Role of Scientific Culture in Modern Society," September 1964

Great American Scientists, Prentice Hall, Editors of Fortune, 1960

Interview of Alvin V. Tollestrup by David A. Valone on December 23, 1994, Caltech Oral History Archives

Interview of Barry C. Barish by Shirley K. Cohen on July 21, 1998, Caltech Oral History Archives

Interview of Barry C. Barish by Shirley K. Cohen, May–July 1998, Caltech Oral History Archives

Interview of Carl Anderson by Harriett Lyle, January 9–February 8, 1979, California Institute of Technology Archives

Interview of Carl Anderson by Harriett Lyle on January 30, 1979, Caltech Oral History Archives

Interview of Carver Mead by Shirley K. Cohen on July 17, 1996, Caltech Oral History Archives

Interview of Charles A. Barnes by Heidi Aspaturian, July–August 1987, Caltech Oral History Archives

Interview with Charles Weiner, March 4, 5 and June 27, 28, 1966; February 4, 1973, Niels Bohr Library and Archives with the Center for the History of Physics

Interview of David S. Wood by Shirley K. Cohen on May 25, 1994, Caltech Oral History Archives

Interview of David S. Wood by Shirley K. Cohen, June 1994, Caltech Oral History Archives

Interview of Fred Anson by Shirley K. Cohen on February 26, 1997, Caltech Oral History Archives

Interview of Hans A. Bethe by Judith R. Goodstein on February 17, 1982, Caltech Oral History Archives

Interview of James A. Westphal by Shirley K. Cohen, July 8–29, 1998, Caltech Oral History Archives

Interview of Jenijoy La Belle by Heidi Aspaturian, February–May 2008, and April 2009, Caltech History Archives

Interview of John Rigden by Dr. Dudley Herschbach on May 21, 2003, American Institute of Physics, College Park, MD, www.aip.org/history/ohist

Interview of John Wheeler by Kenneth W. Ford on March 15, 1994, American Institute of Physics, College Park, MD, www.aip.org/history/ohist

Interview of Joseph Heller by Shelley Erwin on May 5, 2010, Caltech Oral History Archives

Interview of Philip Morrison by Charles Weiner on February 7, 1967, American Institute of Physics, College Park, MD, www.aip.org/history/ohist

Interview of Robert F. Christy by Sara Lippincott, June 1994, Caltech Oral History Archives

Interview of Robert Hellwarth by Joan Bromberg on May 28, 1985, American Institute of Physics, College Park, MD, www.aip.org/history/ohist

Interview of Rodman W. Paul by Carol Bugé on February 17, 1982, Caltech Oral History Archives

Interview of Samuel Epstein by Carol Bugé on December 19 and 26, 1985, and January 10, 1986, Caltech Oral History Archives

Interview of Seymour Benzer by Heidi Aspaturian, September 11, 1990–February 1991, Caltech Oral History Archives

Interview of Steven C. Frautschi by Shirley K. Cohen on June 17, 2003, Caltech History Archives

Interview of Theodore Y. Wu by Shirley K. Cohen, February–March 2002, Caltech Oral History Archives

Interview of Thomas A. Tombrello by Heidi Aspaturian on December 26–31, 2010, Caltech Oral History Archives

Interview of Valentine L. Telegdi by Sara Lippincott on March 4, 9, 2002, Caltech Oral History Archives

Interview for Viewpoint by Bill Stout for KNX Television, ca. 1959

Interview of Werner Heisenberg by Thomas S. Kuhn on February 13, 1963, American Institute of Physics, College Park, MD, www.aip.org/history/ohist

Interview of William A. Fowler by John Greenberg and Carol Bugé, May 1983–May 1984, Caltech Oral History Archives

Julian Schwinger, in his obituary of Feynman in February 1989, *Physics Today*

Les Prix Nobel en 1965 [Nobel Foundation], Stockholm, 1966

"Mass Varying with Position," *Physics* 230, 1987 (R. P. Feynman Papers, California Institute of Technology Archives)

MIT centennial, "Talk of Our Times," December 1961

MIT conference, May 1981; "Simulating Physics with Computers," *International Journal of Theoretical Physics* 21

National Science Teachers Association Fourteenth Convention lecture, "What Is Science?" April 1966

"New Textbooks for the 'New' Mathematics," *Engineering and Science* 28, no. 6, March 1965

New York Times obituary, February 17, 1988

"900 at Caltech, JPL Declare Support for Nuclear Arms Freeze," *The Los Angeles Times*, October 16, 1982

No Ordinary Genius, The Illustrated Richard Feynman, W. W. Norton, edited by Christopher Sykes, 1994

Nobel Lectures, Physics 1963–1970, Elsevier Publishing Company, Amsterdam, 1972

Oersted Medal acceptance speech, 1972

Omni interview, February 1979

Panel discussion, particle physics conference, Irvine, California, 1971

Pasadena Star-News, Opinion, June 18, 1986

People, July 22, 1985

Perfectly Reasonable Deviations from the Beaten Track. Basic Books, Richard P. Feynman, edited by Michelle Feynman, 2005

"The Present Situation in Quantum Electrodynamics," Solvay conference, 1961

"The Problem of Teaching Physics in Latin America," 1963

"The Qualitative Behavior of Yang–Mills Theory in 2+1 Dimensions," January 1981

"QED: Fits of Reflection and Transmission," Sir Douglas Robb Lectures, University of Auckland, 1979

"QED: New Queries," Sir Douglas Robb Lectures, University of Auckland, 1979

"QED: Photons—Corpuscles of Light," Sir Douglas Robb Lectures, University of Auckland, 1979

QED: The Strange Theory of Light and Matter, Princeton University Press, Richard P. Feynman, 1985

"The Relation of Science and Religion," May 1956

"The Remarkable Dr. Feynman," *Los Angeles Times Magazine*, April 20, 1986

Report of the Presidential Commission on the Space Shuttle *Challenger* Accident, Volume 2: Appendix F, June 1986

South Shore Record, October 28, 1965

"Structure of the Proton," Niels Bohr Medal lecture given in Copenhagen, Denmark, October 1973

Surely You're Joking, Mr. Feynman! W. W. Norton, Richard P. Feynman and Ralph Leighton, edited by Edward Hutchings. 1985

Swedish television interview on Nobel Prize winners, 1965

"Theory and Applications of Mercereau's Superconducting Circuits," October 1964

"There's Plenty of Room at the Bottom," December 1959

"Tiny Computers Obeying Quantum Mechanical Laws," *New Directions in Physics: The Los Alamos 40th Anniversary Volume*, 1987

Tom Stoppard, "Stage View," *New York Times*, November 27, 198

UC Berkeley Lectures, "Time and Physics in Evolutionary History," spring 1968

UCSB talk, "Los Alamos from Below," February 1975

"The Uncertainty of Science," John Danz Lecture Series, 1963

"The Uncertainty of Values," John Danz Lecture Series, 1963

Unpublished personal correspondence and notes

"The Unscientific Age," John Danz Lecture Series, 1963

U.S. News and World Report interview, February 1985

"The Value of Science," December 1955

What Do You Care What Other People Think?, W. W. Norton, Richard P. Feynman and Ralph Leighton, 1988

Yorkshire Television program "Take the World from Another Point of View," 1972

Index

Page numbers in *italics* refer to images.